Harold Mytum
Editor

KU-372-567

Global Perspectives on Archaeological Field Schools

Constructions of Knowledge and Experience

LIS LIBRARY

Date	Fund
5-6-13	i -che

Order No
2402245

University of Chester

 Springer

Editor
Harold Mytum
Centre for Manx Studies
School of Archaeology, Classics, and Egyptology
University of Liverpool
Liverpool, UK
h.mytum@liv.ac.uk

ISBN 978-1-4614-0432-3 (Hardcover)   e-ISBN 978-1-4614-0433-0
ISBN 978-1-4614-5964-4 (Softcover)
DOI 10.1007/978-1-4614-0433-0
Springer New York Dordrecht Heidelberg London

Library of Congress Control Number: 2011938860

© Springer Science+Business Media, LLC 2012. First softcover printing 2012.
All rights reserved. This work may not be translated or copied in whole or in part without the written permission of the publisher (Springer Science+Business Media, LLC, 233 Spring Street, New York, NY 10013, USA), except for brief excerpts in connection with reviews or scholarly analysis. Use in connection with any form of information storage and retrieval, electronic adaptation, computer software, or by similar or dissimilar methodology now known or hereafter developed is forbidden.
The use in this publication of trade names, trademarks, service marks, and similar terms, even if they are not identified as such, is not to be taken as an expression of opinion as to whether or not they are subject to proprietary rights.

Printed on acid-free paper

Springer is part of Springer Science+Business Media (www.springer.com)

Contents

Part III   Underwater

Part IV   Non-Excavation

Part V   Fieldwork and People

Contributors

Ran Boytner Institute for Field Research, Department of Anthropology, University Park Campus, University of Southern California, Los Angeles, CA, USA

John R. Bratten Department of Anthropology, University of West Florida, Pensacola, FL, USA

Alasdair Brooks School of Archaeology and Ancient History, University of Leicester, Leicester, UK

Bonnie J. Clark Department of Anthropology, University of Denver, Denver, CO, USA

Amanda Clarke Department of Archaeology, University of Reading, Reading, UK

Hannah Cobb Archaeology, School of Arts, Histories and Cultures, University of Manchester, Manchester, UK

Sarah Colley Department of Archaeology, School of Philosophical and Historical Inquiry, University of Sydney, Sydney, NSW, Australia

Karina Croucher Archaeology, School of Arts, Histories and Cultures, University of Manchester, Manchester, UK

Robert I. Curtis Department of Classics, University of Georgia, Athens, GA, USA

Anne Corscadden Knox PAST Foundation, Columbus, OH, USA

Benedict Lowe Department of History and Area Studies, University of Aarhus, Aarhus C, DK, Denmark

Betty Jo Mayeske History and Humanities, University of Maryland University College, Adelphi, MD, USA

Harold Mytum Centre for Manx Studies, School of Archaeology, Classics, and Egyptology, University of Liverpool, Liverpool, UK

Tim Phillips Department of Archaeology, University of Reading, Reading, UK

Timothy James Scarlett Department of Social Sciences, Michigan Technological University, Houghton, MI, USA

Sheli O. Smith PAST Foundation, Columbus, OH, USA

Sam R. Sweitz Department of Social Sciences, Michigan Technological University, Houghton, MI, USA

Lisa C. Young University of Michigan, Museum of Anthropology, Ann Arbor, MI, USA

Part I
Theory and Practice

Chapter 1
Introduction: Constructing Education and Knowledge in the Field

Harold Mytum

1.1 Teaching and Researching Archaeology at University Level

The teaching of archaeology is of central concern to students, staff, and the wider community of heritage professionals for whom some at least of the graduates will work (RPA 2003). It is also important that the training includes the acquisition of sufficient generic skills that all graduates will be equipped to find suitable employment in the wider economy.

In recent years, a substantial literature on pedagogic theory has accumulated on how University staff should provide theoretically and methodologically informed courses for their students (Cown 2006; Fry et al. 2009; Martinez-Pons 2003), and some of this literature also applies to those with lesser responsibility (Ward et al. 1997). The growth of training and the establishment of professional qualifications across the globe have led to a greater awareness of both theory and practice of teaching in general and its particular application within higher education. The Certificate in Professional Studies in Learning and Teaching is a compulsory qualification for early years UK faculty to gain during their probationary period in their posts, and similar frameworks exist elsewhere. Reflexivity in both teaching and learning is now a major emphasis (Chap. 2), and this book is an example of this.

Concern with the teaching of archaeology and the role of universities within the wider professional heritage framework have been issues under active debate within the profession for over a decade, not only in North America (Bender and Smith 2000; Derry and Malloy 2003; Jameson and Baugher 2007) but also elsewhere (Aitchison 2004; Brookes 2008; Colley 2004; Darvill 2008; Rainbird and Hamilakis 2001; Ucko et al. 2007; Ulm et al. 2005). However, most of that which has been

H. Mytum (✉)
Centre for Manx Studies, School of Archaeology, Classics, and Egyptology,
University of Liverpool, Liverpool, UK
e-mail: h.mytum@liv.ac.uk

H. Mytum (ed.), *Global Perspectives on Archaeological Field Schools:*
Constructions of Knowledge and Experience, DOI 10.1007/978-1-4614-0433-0_1,
© Springer Science+Business Media, LLC 2012

published has been brief summaries of conference papers or round table discussions, and many papers have been linked to specific issues that were important at a particular time or place. Moreover, only some of this discussion is directly relevant to field schools, though it may provide the frameworks in terms of curricula within which field training takes place, or the form and content of classroom teaching which forms the basis of student knowledge prior to their field experience, and to which they subsequently return and apply their enhanced and expanded understanding within their continuing education, and indeed in their work lives thereafter. Baxter (2009) has provided a practical guide to planning and running many aspects of a North American field school, and this volume does not intend to repeat much that is outlined there. Rather, here a more comparative, international perspective, also informed by recent research into field training in other disciplines (Fuller et al. 2010), augments and expands the pedagogic and research frameworks of field schools and reveals the many positive benefits of a field school experience.

1.2   The Purpose of This Volume

I remember giving a paper at an SHA conference in which I discussed the feelings, emotions, planning, and operation of a particular field school, discussing reflexively many of the issues that experienced field school directors would recognize, though we rarely share in a coherent manner. As the session finished, a student next to me who was just starting her first year of graduate study commented that she had never realized how much thought went into structuring and organizing an excavation with students, and how many factors had to be taken into account and dealt with as the project developed. The case studies and the accumulated experience – perhaps even in places wisdom – that is shared in these chapters should also help students appreciate how they form part of this complex endeavor that comprises a field school (Chap. 15).

   This book is designed to help all those involved in the particular endeavor of field school teaching and learning to gain most from the contributors' personal experiences. All of us started as novices, and we are now at varied stages of our careers, having progressed through levels of responsibility to now manage projects that combine teaching and learning with primary research (Chap. 13). The chapters, all written by experienced field school directors, examine particular emphases and approaches, lessons, and warnings that are important for all those involved in field teaching. At first reading, the chapters may seem more relevant to those planning to direct their first field school, or those wishing to develop new aspects of a field school program, and this is certainly one intention. The book, however, is also to be of value to potential students considering attending their first field school, and graduate students who have been asked to assist as a junior staff member on a field school team. Discussions of pedagogy should be of interest to the taught as well as the teachers (Chap. 2).

Many of the issues that apply to archaeological field schools may have been considered more deeply in some parts of our discipline than others, such as health and safety in underwater programs (Chap. 9) and industrial site archaeology (Chap. 8), but these issues relate more widely, as the referenced literature in these chapters indicates. It is possible to learn from the experiences of the numerous sub-fields within our discipline, and a novice field school director who just looks at field schools superficially similar to the one that is planned can limit their vision of the possible; for those of us more experienced, ignoring others' innovations in teaching, project design or implementation can mean that we do not learn from others' challenges. For example, project organization has been a high priority because of the expense and complexity of underwater archaeology (Chap. 10), but many issues of balancing various priorities and having alternative schedules apply to all excavation and survey projects (Chaps. 7 and 11). All field schools require large amounts of planning, and although some items can be derived from check lists devised by others (Baxter 2009), no two projects are the same and nothing should be merely copied as this line of least resistance at this stage will lead to logistical problems at a later stage.

The aims and objectives of field school training vary not only between subdisciplines, but also across countries, with their different higher education structures (Beck 2008; Chap. 5) and their varied archaeological historiographies. This is reflected in the amount and content of typical field training, and the proportion of students expected to take part, with all archaeology graduates having some exposure to practical field archaeology in the UK (Chap. 3), through a small proportion of those taking archaeology field courses in North America (Chap. 6), to no training offered at all at some Australian universities (Chap. 5). More is not necessarily better; what matters for those who attend is the quality of their experience (Thorpe 2004) and its applicability later in their careers. The treatment of similar finds can vary across the globe (Chap. 12), as can the attitudes to the acquisition of transferrable skills (Chap. 3) or enabling those with a disability to participate (Chap. 4). These variations need to be considered when students from one archaeological tradition venture out to experience another; this may apply to prehistorians working on a historic site, terrestrial archaeologists going underwater (Chap. 9), or North Americans venturing to the UK or Europe (Chaps. 7 and 11).

Education in archaeological field practice does not only come through formal field schools, the main focus of study here, but also through community archaeology in many forms (e.g. Cressy et al. 2003; Jameson and Baugher 2007). The combination of a field school with community involvement, and understanding that community's understanding of its heritage, is central to some field schools (Silliman 2008; Chap. 14), but it is also present in many others (Chaps. 8 and 10). While a mix of university students and other learners is more common in British and some European terrestrial archaeology field projects (Chaps. 4, 7, and 11), it is also common in underwater archaeology where there is a strong desire to inform and educate already experienced divers about the value of the archaeological heritage and the need to protect it (Chaps. 9 and 10). The central role of nonarchaeologists in industrial archaeology, whether they be engineers, historians, or architects, is well known, and this tradition is also maintained in North America (Chap. 8). In all these projects, a more varied clientele creates challenges in formal training because students do not share a base of common prior experience, but also opportunities for students to learn much from each other.

One of the most important but least often acknowledged advantages of the field school is the way in which students learn skills that are valuable in all forms of employment. Research in the UK has revealed how rarely students, and indeed many staff, reflect upon the transferrable skills that field schools encourage and develop (Chap. 3). Most North American literature concentrates on the skills required for professional development, perhaps because a much smaller proportion of students subscribe to a field school experience and so may already be considered self-selected as potential archaeologists. However, many will not succeed in the long term in this most competitive and often poorly paid of professions, and the skills that students and junior project staff learn should be explicitly acknowledged.

All field schools involve everyone – staff and students alike – learning from the experience (Walker and Saitta 2002). While a book cannot encapsulate all that is experienced in a field school, this mix of analytical summaries in the *Theory and Practice* section, combined with the remaining nine chapters from a variety of case study perspectives, can provide advice and inspiration for aspiring students, assistant staff, new field school directors, and old hands at every level to consider how, where, and why they wish to spend their time on a field school and to help them get the most out of that most iconic of experiences.

Acknowledgments I would like to thank all the authors for their interest and some for their patience as this volume has been compiled. It is a tribute to their dedication to this most important aspect of archaeological education that they have provided these reflective studies to help others achieve their goals. Teresa Krauss and Morgan Ryan at Springer have been helpful and encouraging at every stage and helped keep the momentum going as we all fitted in producing our contributions between completing one field school and planning the next. I would also like to thank all those who have taught me so much about field archaeology and teaching; from my first excavation directors and supervisors to my most recent cohort of field school students, I have learnt from you all.

References

Aitchison, K. (2004). Supply, demand and a failure of understanding: Addressing the culture clash between archaeologists' expectations for training and employment in "academia" versus "practice". *World Archaeology, 36*(2), 203–219.

Baxter, J. E. (2009). *Archaeological field schools. A guide for teaching in the field*. Walnut Creek: Left Coast Press.

Beck, W. (2008). *By degrees: Benchmarking archaeology degrees in Australian universities*. Armidale, New South Wales: Teaching and Learning Centre, University of New England.

Bender, S. J., & Smith, G. S. (Eds.). (2000). *Teaching archaeology in the twenty-first century*. Society for American Archaeology, Washington.

Brookes, S. (2008). Archaeology in the field: Enhancing the role of fieldwork training and teaching. *Research in Archaeological Education, 1*. Retrieved March 21, 2011, from http://www. heacademy.ac.uk/hca/archaeology/RAEJournal

Colley, S. (2004). University-based archaeology teaching and learning and professionalism in Australia. *World Archaeology, 36*(2), 189–202.

Cown, J. (2006). *On becoming an innovative university teacher. Reflection in action* (2nd ed.). Maidenhead: Society for Research into Higher Education and the Open University.

Cressy, P. J., Reeder, R., & Bryson, J. (2003). Held in trust. In L. Derry & M. Malloy (Eds.), (2003), *Archaeologists and local communities: partners in exploring the past* (pp. 1–17). Society for American Archaeology, Washington.

Darvill, T. (2008). UK Archaeology benchmark updated. *Research in Archaeological Education, 1*. Retrieved March 21, 2011, from http://www.heacademy.ac.uk/hca/archaeology/RAEJournal

Derry, L., & Malloy, M. (Eds.). (2003). *Archaeologists and local communities: Partners in exploring the past*. Society for American Archaeology, Washington.

Fry, H., Ketteridge, S., & Marshall, S. (Eds.). (2009). *A handbook for teaching and learning in higher education*. New York: Routledge.

Fuller, I., Brook, M., & Holt, K. (2010). Linking teaching and research in undergraduate physical geography papers: The role of fieldwork. *New Zealand Geographer, 66*, 196–202.

Jameson, J. Jr., & Baugher, S. (Eds.). (2007). *Past meets present. Archaeologists partnering with museum curators, teachers, and community groups*. Springer, New York.

Martinez-Pons, M. (2003). *The continuum guide to successful teaching in higher education*. London: Continuum.

Rainbird, P., & Hamilakis Y. (Eds.). (2001). *Interrogating pedagogies: Archaeology in higher education*. British Archaeological Reports International series 948, Oxford.

RPA; Register for Professional Archaeologists. (2003). Guidelines and standards for archaeological field schools. Electronic document. Retrieved October 13, 2010, from http://www.rpanet.org/associations/8360/files/field_school_guidelines.pdf

Silliman, S. W. (Ed.). (2008). *Collaborating at the trowel's edge: Teaching and learning in indigenous archaeology*. University of Arizona Press, Tucson.

Thorpe, N. (2004). *Student self-evaluation in archaeological fieldwork*. Report for the Higher Education Academy. Retrieved March 21, 2011, from http://www.heacademy.ac.uk/hca/resources/detail/student_self_evaluation_in_archaeological_fieldwork

Ucko, P. J., Ling, Q., & Huber, J. (Eds.). (2007). *From concepts of the past to practical strategies. The teaching of archaeological field techniques*. London: Saffron.

Ulm, S., Nichols, S., & Dalley, C. (2005). Mapping the shape of contemporary Australian archaeology: Implications for archaeology teaching and learning. *Australian Archaeology, 61*, 11–23.

Walker, M., & Saitta, D. J. (2002). Teaching the craft of archaeology: Theory, practice and the field school. *International Journal of Historical Archaeology, 6*(3), 199–207.

Ward, A., Baume, D., & Baume, C. (1997). *Learning to teach: Assisting with laboratory work and field trips; training materials for research students*. Oxford: Oxford Centre for Staff Development.

Chapter 2
The Pedagogic Value of Field Schools: Some Frameworks

Harold Mytum

2.1 Introduction

In North America, the field school is often described as the rite of passage during which those students who will develop an archaeological career discover what the emblemic activity of excavation involves and identify themselves as aspiring archaeologists (Baxter 2009; Chaps. 6 and 13; Walker and Saitta 2002). In the UK, less formalised training excavations are available to those at school as well as university and are well-advertised through media such as the Council for British Archaeology's website. In the UK, more opportunities arise for other forms of involvement because of the traditions of avocational archaeology carrying out scientific excavations and because of community involvement sponsored by the Heritage Lottery Fund, but in other countries such as many in Europe and in Australia (see Chap. 5) the local opportunities are often extremely limited. Nevertheless, in most countries it would seem that the formal field training offered to students while undergraduates at University is the recognised building block on which all further professional and academic experience is founded.

While Baxter (2009) has produced a valuable outline for the aspiring field school director in North America, a single-author volume is by necessity limited in the range of first-hand experiences that can be used to illustrate problems and challenges and their solutions. Issues surrounding projects intimately linked with indigenous communities have been explored through the volume edited by Silliman (2008), though again only in North American contexts. Field training can be extremely variable, from an element in a primarily research-driven project (see Chap. 11), to one with a mix of attendees including those taking formal classes

H. Mytum (✉)
Centre for Manx Studies, School of Archaeology, Classics, and Egyptology,
University of Liverpool, Liverpool, UK
e-mail: h.mytum@liv.ac.uk

H. Mytum (ed.), *Global Perspectives on Archaeological Field Schools:*
Constructions of Knowledge and Experience, DOI 10.1007/978-1-4614-0433-0_2,
© Springer Science+Business Media, LLC 2012

for credit and university students obtaining experience as part of their degree programme that is not formally assessed (as has often been the case in the UK), through to volunteers of any age who wish to gain an understanding of archaeology (see Chap. 7), including those with various forms of disability (see Chap. 4).

The North American field school pattern of 4–6 solid weeks of fieldwork is only one model for the instruction of archaeological practical methods. In the UK, more rigorous degree programme planning has formalised teaching methods, and these are now often structured on a developmental basis with short (5–14 day) compulsory intensive programmes where teaching dominates over research for all first-year archaeology students, with then optional modules in the second year where greater depth of training is offered, often linked to staff research projects (see Chap. 3). Whereas the field school is a discretionary element in North America and may be only available at a few institutions in Australia (see Chap. 5), it is seen as central in the UK where archaeology is recognised as a discrete discipline, even where it has been combined in larger units within university institutional frameworks. Fieldwork experience is part of the national subject teaching framework (Darvill 2008), and departments have to design some way of fulfilling this requirement. This nationally recognised status for fieldwork training has assisted departments in protecting internal budgets for this activity in the face of financial constraints, but it is still a major logistical challenge for most if not all departments, particularly as cohort sizes have increased substantially. However, as most students do not wish to continue in the archaeology profession and the fieldwork requirement beyond a basic level is not compulsory, it is less problematic to find resources to support the relatively small numbers who continue with this field training through their degree programmes. This different structure, combined with many ways in which an extremely visible heritage can be experienced by students other than on projects, means that the field school experience does not hold quite the same pivotal place in the emotions of many UK students and staff.

Whatever the overall structure and timing of practical archaeology, it is fieldwork that converts many students to the subject, just as it puts others off this as a career. Therefore, how excavation and survey is projected, taught, and learnt in field schools affects how the profession will consider its role in the future. The widespread imbalance by gender towards males in professional field archaeology and to females in field schools for North American and UK students (see Chaps. 4 and 6), yet the high proportion of museums and some other heritage sector posts being now held by women, suggests that the significance of the traditional field school in creating an interest in the wider heritage profession may be outdated if it concentrates on excavation. In this regard, field school training on finds (see Chap. 12) or programmes where survey or public interaction is important (see Chaps. 11 and 14) may be more significant than has hitherto been appreciated.

The field school is designed to meet a number of training aims that are achieved through a series of objectives that the director structures within the time students spend on the project. There are two aspects that uniquely define almost all field schools: that experiential learning is central, and that learning takes place within a real-life research context. Each of these require careful consideration, as much of the generic literature on student learning assumes a primacy of classroom contexts and a control and predictability of student experience which is not always applicable in a field context.

2.2   Experiential Learning

Experiential learning is one of the key features of the field school; students may have lectures and other forms of formal instruction, but learning through doing is central to its ethos. It is therefore worthwhile considering the theoretical basic of this approach, but also what has been learnt from cognate disciplines such as geography and geology that also incorporate practical expeditions within their programmes (Healey 2005; Fuller et al. 2010; Spronken-Smith and Hilton 2009). Indeed, there has been a longer tradition of studying the pedagogy of fieldwork in those disciplines than there has in archaeology, so it is possible to benefit from these insights in considering both ways in which students can learn within fieldwork and how they might produce assignments for assessment.

Learning is seen by Kolb (1984) as a cyclical process in which concrete experience leads to reflective observation (Fig. 2.1). This is followed by abstract conceptualisation that has an impact through active experimentation, which in itself leads to new concrete experiences. The concrete experience can be considered as that gained in any number of field school activities, but within this the reflective observation can be of several kinds. It may be on the student performance – what am I actually doing within this task? But it can also be recalling reflective observation of past concrete experiences earlier in the field school or from other fieldwork, or may be reflecting on the form of recovered data in the ground and within the created record.

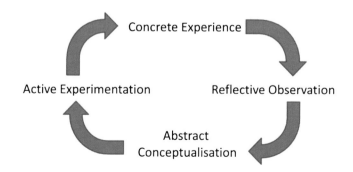

Fig. 2.1   The learning cycle according to Kolb (1984)

From this follows abstract conceptualisation, and this may relate to questions such as what does this mean, or what could this data be used for, consideration of how a post-mold represents an above-ground feature on its own, and perhaps part of a more complex structure. Issues of sequence and phases may be more apparent as students elucidate the stratigraphic relationships between features or while drawing a profile, and social or cultural inferences may be more obviously inspired by working with artefacts. In the Kolb cycle, this then leads on to active experimentation, for example, adjusting the excavation or recording techniques slightly, wondering about alternative interpretations.

Kolb (1984) considered that learners have one of a variety of learning styles, and this approach has been applied to the teaching of geography fieldwork at university, not dissimilar to many archaeological situations:

> *Divergers* view situations from many perspectives and rely heavily upon brainstorming and generation of ideas.
> *Assimilators* use inductive reasoning and have the ability to create theoretical models.
> *Convergers* rely heavily on hypothetical-deductive reasoning.
> *Accommodators* carry out plans and experiments and adapt to immediate circumstances (Healey and Jenkins 2000).

It is clear that explaining to students, structuring group and individual learning, and appreciating and empathising with the format and emphasis of the field school research design, will not be the same for all groups. Indeed, the way in which the field school is described and defined might influence which types of learners select a particular field school. While geographers have considered learning styles in relation to their students (Marvell 2008), this has yet to be attempted in archaeology.

While all learners will experience all stages of the Kolb cycle, Honey and Mumford (1982) have argued for a different categorisation of learning styles in which each of the defined types of student engages more strongly with different parts of Kolb's cycle. They suggest that learners may be divided into activists, reflectors, theorists and pragmatists. Moreover, they may prefer diverse forms of exploring the experiences and issues raised while learning. Activists not surprisingly focus on the concrete experience, and so these can be among the most motivated in the actual process of fieldwork, but may be weaker when it comes to reflecting on why decisions have been made the way they have, or in their flexibility as situations change. The reflector enjoys making observations and considering their implications, and so some forms of recording would be most likely to appeal to this type of student. Theorists may be less interested in the practical activity, seeing it only as a means by which to reach deeper understanding; such students can be poor on detail and untidy, irritated at the slow progress of fieldwork and the unprocessed nature of the data which prevents immediate interpretation. They may become time-wasters asking too many questions and engaging in discussions which may be appropriate in class but not in a practical work situation. The pragmatist, in contrast, wants to apply insights already gained in preliminary field courses and reading or earlier in the field season to the newly available practical situations, and abstract issues such as theory, regional methodological traditions, and project research goals may be seen as impediments to getting on with the work.

The reflectors and theorists can be most effective in traditional university courses, but may do less well in the field context. They can find the practical elements, including organisation involving others, the need to carry out repetitive tasks, and the response that we do not yet know the answers to their high level questions, all extremely frustrating. In contrast, activists and pragmatists can find that they are in their element, being able to carry out clearly defined programmes of action toward defined goals. Fleming and Baume (2006) have proposed a different set of categories for student learning styles, emphasising visual, aural, reading and writing, or kinaesthetic styles, and these likewise carry the implication that some students may be generally stronger in the field school context than in class, and vice versa, and that some students will absorb some certain of communication and learning more easily than others.

Types of knowledge also appeal to different learning styles, and this also needs to be considered in the content and structure of a course. Convergent knowledge combines principles with particular information to deal with a problem, as with understanding the principles of stratigraphy on the one hand combined with the drawing and interpretation of an actual profile through the construction of a Harris matrix on the other. This type of knowledge usually leads to right or wrong answers – the sequence is either correctly identified or it is not. Divergent knowledge links to experience and judgement and relates to qualitative matters that do not have clear right or wrong solutions, as with decisions regarding sampling deposits for environmental data. In fieldwork contexts, some yearn for a firm right or wrong answer, others relish the challenges of shades of grey rather than black or white; this will affect what is seen as a satisfactory conclusion to a discussion and can vary from one cohort to another. For some, a definite answer is required to make a learning session successful, for others revealing its complexity is the desirable outcome. The field school teacher needs to be ready to help all to their own appropriate outcomes, rather than necessarily to a full consensus. Evaluating students and their personalities, including learning styles, is an important part of the field school staff's role and should commence as soon as the project begins.

One of the criticisms of the Kolb model is that learning does not take place in the straightforward cyclical order of the model, and this is very obvious when carrying out fieldwork. There is a constant interaction between several of these stages at any one time, the feedback being much more fluid than is implied by the diagram. Moreover, the rate and quality of learning and reflection varies over time, depending on the task, the student's motivation, and the levels of new information (about the task or the archaeology) that is being discovered. It is important for both students and staff to be aware that all stages of this model should be stimulated and encouraged. Just "getting the job done" on site is insufficient, but so is discussing abstruse points of theory or method without achieving any practical output, not only in terms of the research aims but also pedagogically, given the practice elements of the aims and objectives.

2.3 Experience in a Research Context

Experiential learning can occur to varying degrees away from the field, such as artefact or ecofact practical classes in laboratories, or excavation research design testing through computer simulations. What makes the field school special for both staff and students is that the teaching and learning happens partly alongside but largely within a research context. Few academic subjects can allow students to experience a class where they can be contributing to cutting edge research while also learning. Some students find it hard to appreciate that their professor does not know all the answers – the response "If we knew all that already we would not need to excavate" is unexpected. But the very engagement with current research questions, however small-scale, creates both opportunities and practical and ethical concerns in relation to the pedagogic aims of the field school (see Chaps. 8, 10 and 14). Students must be aware that field schools have, in their very operation, commitments to them as learners, but also to the public, the wider archaeological community, and to the discipline itself (see Chap. 15).

The design of teaching and learning activities has to be matched with practical and ethical issues, especially where these involve excavation and thus the destruction of the archaeological resource (Fig. 2.2). In many cases, these issues are predictable and the explanation of these to the students before and during work makes them both appreciate the importance of their contribution, and its quality, and is part of the student learning experience about what archaeological fieldwork involves. The challenge comes, however, when the archaeology is not as predicted. If the range of materials and features, or the density of artefacts recovered, is far less than planned, thus limiting the expected student experience, alternative strategies may need to be employed to complete learning objectives even if these are not part of the research design. In contrast, the data may be far greater and more complex than expected, creating challenges for inexperienced excavators and recorders, and for the project infrastructure; managing the material may dominate staff attention over teaching students the methods. More than with other forms of teaching, the director and other staff need to be more adaptable and ideally have alternative strategies ready to apply in such situations. One of greatest risks for the research effectiveness of a project is that the field training appears to have gone well, but the records, artefacts and samples have not received the attention they deserved and there is a large backlog of work to be undertaken even though most of not all the labour will have departed (Baxter 2009). Training needs to be structured to give appropriate emphasis to all stages of the archaeological process, and support staff need to be of an experience and number to cope with all the necessary checking and ordering of records and artefacts as the project proceeds.

There is no doubt that one of the greatest benefits to students is their increased motivation engendered by their involvement with discovery. At one level this is personal – finding a shard of pottery, a wall, discerning a change in colour or texture of the soil. But at another it is that such discoveries, and those made together as a team with research goals, lead to understanding things about the past that were not known when the field school began. It is therefore necessary for the field school director to build into the aims and objectives the learning about the contingency of

Fig. 2.2 Teaching takes place in various formats, involving action and problem-solving in both excavation and recording

research, of the messy nature of real-life archaeological data, of the provisional nature of interpretations and predictions, and that negative as well as possible results all contribute to understanding. Field archaeology should encourage a reflexive approach by students (Cobb and Richardson 2008; Hamilakis 2004), just as it must for staff in both their teaching and research.

2.4   Syllabus Content and Assessment

Field schools vary, and while the majority incorporate excavation, whether terrestrial or underwater, some focus on survey (Chap. 11) or on finds or conservation (Mills et al. 2008). There cannot be one overall approved syllabus, but it is important that each director determines what aspects of archaeology are to be formally taught, what aspects students may become optionally exposed to or have greater focus upon, and what informal types of learning are to be specifically encouraged by exposure through site visits, meetings with other stakeholders in the heritage, or interaction within the field school group.

Professional archaeological bodies such as the Institute for Field Archaeologists (IfA) in Britain or the Register of Professional Archaeologists (RPA) in the US have particular defined skills which are seen as essential for all those entering the profession (Adler 2001; Perry 2004). Most academics would accept these as ones that could be taught within a field school, but not all posts in CRM or other professional structures require the same skills (Schuldenrein and Altschul 2000). It is therefore necessary to consider to what level of proficiency in any skill a student should reach; it is hard to find the time necessary for students to gain experience to a level that can be applied professionally (Neusius 2009). To have assisted in carrying out a gradiometer survey is quite different from having the responsibility of undertaking one in the field, downloading the data, manipulating it effectively, and interpreting it. Likewise, excavating in a test pit in the supportive and relatively slow working environment of the field school may or may not prepare a student for rapidly excavating a number of test pits within a CRM context. There are differences between some awareness of a technique, experience of it, competence in a supervised environment, and independently capable. These gradations should be made clear in the learning objectives – what level for a task will students be expected to reach? The challenge with meeting the expectations of those in the profession is that when starting from absolutely no practical knowledge, even 6 weeks is a short time to gain proficiency in many skills. Moreover, much professional experience is gained by working on different sites and in different conditions, something not possible for a single field school. It is also important that students' expectations are realistic regarding how much they can gain from one field school. Some students will be naturally able and will advance quickly, others may find even the most basic tasks a challenge. While a few participants will have combined attitude and abilities that allow them to reach a significant level of competence, learning outcomes need to be set so that most can achieve a recognised, albeit lower, level. We do not expect students to produce publication-quality written work to pass, and the same equivalence needs to be applied to practice; as field school grades are rarely based on practical skills, the profession needs to appreciate the role and limitations of field schools within the university structures that exist in each country.

The structure of the syllabus is partly constrained by the form of a field season. On a site with no already partially excavated areas, work of necessity begins with setting out excavation areas and the removal of the plow zone. Only once profiles have been revealed can their significance be explained and their recording proceed. Although preliminary lectures may set the scene on recording methods to be employed, the experiential learning only takes place as these stages of the excavation are reached. This is both a constraint but also an opportunity, as the very sequence of an excavation is in and of itself something that students should appreciate. Some activities can begin at quite an early stage, for example artefact processing even from plow zone collections, but some of the issues regarding use of artefacts for dating and functional interpretation may have to await the discovery of intact deposits later in the season. Indeed, learning the use of excavation tools and the procedures for artefact processing on superficial deposits of potentially less research value can be a less stressful way for both students and staff as the initial training takes place. Just as students are generally unused to the concepts of industrial labour (see Chap. 8), so many are unfamiliar with hard physical effort, and so fitness levels are sometimes problematic; while some activities may be constrained by disability (Chap. 4), many students are not as fully able to sustain physical effort as they were 20 years ago. Gradually gaining fitness is one of the unexpected by-products of a field school for many students.

In many field schools not all students work on a particular activity at one time, but rather these are linked to the necessary stages that their part of the excavation has reached. It is important to have some form of ongoing record that keeps track of which student has had which experience and the level of training and consolidation in understanding that they have received. It is easy for some aspects of training to be postponed, especially if labour-intensive for staff and requiring small group teaching, or limited to few locations on the site. This will lead to logistical problems towards the end of the field school, at a time when the completion of research priorities and the checking of the site archive are also at its height. It also means that issues introduced towards the end of the field school have limited time for application and integration in the students' minds.

Assessment of fieldwork can be a major concern, both for staff and students. While all are familiar with the criteria that apply to standard modes of class assessment, they may not be relevant or most appropriate in the field school context. It is therefore particularly important that explicit guidelines are drawn up by staff, and that grade descriptors are available to students as well as those responsible for grading. The grading may have to be broken down into various elements, suitably weighted, as students may for reasons discussed above be variable in their effectiveness across the spectrum of tasks involved. A useful summary of the principles, with many examples of grade descriptors, is provided by Prosser (2010), though most are designed for written assignments. These tabulations and correlations may be particularly valuable if students come from more than one institution, and perhaps other countries. As students progress through various activities, some method of recording levels of attainment needs to be instituted so that final grading is not based on only the most recent activities.

Some field schools require written assignments (see Chap. 7) and others incorporate a portfolio of records and a diary, while others rely on assessment of actual practice. There are good arguments for including practical competence, but also raise issues of measurement and comparability (one student's test pit may be complex, another's sterile). Whatever the balance, this needs to have been thought through and integrated into the aims and objectives of the programme and can include written or oral, the latter most appropriate in a fieldwork setting where students are taking some level of responsibility, for example in test pits; in geography this form of field assessment has been shown to be very successful (Marvell 2008).

Modes and timing of assessment should also be reviewed through some form of student feedback. There may be arguments for a brief survey before the field school begins and another afterwards, to see if perceptions change and expectations are met; studies of geography students show positive shifts between pre- and post-field trip cohorts (Boyle et al. 2007), and for students with disability, this reflexivity has assisted both students and staff (see Chap. 4); less complex versions for all students could be valuable. Most students, and indeed staff, do not recognise or explicitly state the transferable skills gained during the field school and its assessment (Chap. 3, Table 1). More emphasis should be placed on these strengths, which may also form part of the assessment.

2.5   Managing the Risks

Field schools provide an educational experience where there is a mix between formal teaching, backed up by reading, with a relatively large amount of time devoted to practical activity. While this resembles laboratory practical work in some respects, and so is similar to the pedagogic experience in other aspects of archaeology and some science subjects, it differs in some important respects. Except for a small number of excavation simulations, most field school fieldwork takes place in a context where not all variables are under the instructor's control. Unlike classroom teaching, contact times are long, and many factors can affect activity and learning. Moreover, although most research students and junior faculty are now provided with elaborate and sometimes time-consuming training in pedagogic skills in many countries, these cover standard contact situations such as one-to-one tutorials, small group teaching in seminars and workshops, lectures and laboratory practicals. Even where outside project work is occasionally considered, this is envisaged as being very controlled local repeatable simulations of site visiting or data collection that are still largely unproblematic.

Consideration of risk normally concentrates on health and safety, and all field school directors should be trained in evaluating and having strategies for all risks that they can perceive for their project. Universities all have administrative procedures to ensure that their insurance terms are covered, an issue that can cause problems when students attend projects runs by other institutions (see Chaps. 5 and 6). It is essential that, however bureaucratic these systems may be, all key staff are fully trained and aware of all these procedures and limitations on activity. It is also highly desirable to have a number of staff trained on first aid at the appropriate level, with greater training and qualification necessary for those expeditions to remote areas. It is not sufficient to only have the field school director trained, as they may have to be away from the site or base camp for administrative reasons at times, and also someone has to be able to deal with the director's injury should they suffer one. Student and junior staff awareness of the known abilities and disabilities of attendant students is an essential element in preparation and planning (see Chap. 13), even at the level of acceptance and adding of conditions and provisos for those who may not be able to fully achieve all the standard learning objectives or may require a different form of assessment. Even minor issues such as colour blindness will affect student perception of deposits or artefact characteristics and the extent to which they can complete recording protocols successfully.

The awareness of health and safety constraints affects how teaching will take place, and where; it may affect group size, proportions of staff to students, and where students might work such that they could be evacuated in an emergency. While some conditions are obvious because they are frequently an issue, such as trench depth and the risk of collapse, others such as proximity to a rock face, unstable structure, or a busy road may not be. Field school directors draw from their own personal experience, but in health and safety many issues are also localised and require a level of anticipation beyond direct experience. This is particularly important when training students as, unlike a professional field crew, they are often unaware of the dangers of earthmoving machinery, wildlife, or the many forms of trip hazard, to name just some examples. Discussing the possible hazards with colleagues and more experienced project directors is an excellent way to increase awareness; those many stories told in bars at conferences of heroic acts and "near misses" can actually provide a framework of warnings that should be heeded. These stories are not desired outcomes from a field school, and all field staff should be constantly vigilant regarding student action or lack of it with regard to these matters. The learnt awareness of health and safety is itself part of the field training and is indeed an excellent transferable skill that can be applied in other work contexts. It must be a condition of attendance that all students abide by all regulations, and that refusal to wear, say, stout shoes or a safety helmet when required should exclude them from the site. Any toleration of lax standards will only lead to increased risk for which the director will be held responsible, but also encourage lower quality in other aspects of the project such as excavation and recording. While as director these rules may seem a hindrance, they should be seen and broadcast as the enabling framework within which work can take place at all. Particular risks can be attendant on work in industrial archaeology (Chap. 8), and even more obviously with underwater archaeology (Chaps. 9 and 10).

The logistical uncertainties beyond health and safety also create a sense of tension, particularly in the minds of inexperienced project directors, and these are always a matter of concern, especially on the first field season on a site. The nature of the buried deposits, their state of preservation, the range and density of artefacts, and the period uses of the site are all archaeological variables that affect training, research design, and allocation of resources to different aspects of the project. To these archaeological unknowns can be added relationships with neighbouring contemporary communities (including the landowner), social dynamics within the students and staff, variations in weather, reliability of equipment including transport, and unforeseen problems with accommodation and food. While on-campus practicals can face some of these problems, there is locally available support, alternative facilities, or at worst the re-scheduling of relatively short teaching sessions. Out in the field, some of these problems can be difficult to resolve and can severely impede or even halt all work, and so affect the pedagogic value of the experience for students.

Fig. 2.3 Risk management: wet-weather activities designed to incorporate artefact and sample processing when on-site activities were curtailed, ensuring that teaching and learning could continue in a flexible manner

Students like time off, but they do not appreciate missing out on the training for which they have given both time and money. Many logistical problems are outside the control of the field school director, but that is not appreciated by inexperienced students, who see themselves as customers who were promised a particular product. It is therefore important to incorporate flexibility in perhaps the order that training in particular aspects may take place, or the replacing of one type of practical activity with another, with perhaps merely discussion and reading representing the lost activity. It is valuable to have a number of exercises and projects, as well as background reading, held in reserve to cover at least the first stages of any crisis. This can allow junior staff to supervise students on this activity while the director or the relevant staff member can concentrate on resolving the problem, or determining the best course of action if a solution is not immediately to hand.

Examples of a flexible approach can best explain how field school directors can be prepared for difficulties. In the UK, it can rain heavily for several days in a row and, while some work in the rain may be possible, on many subsoils the site rapidly becomes unworkable, and damage is done to deposits by attempting excavation and recording. The Castell Henllys base camp included a large dining tent, and after breakfast on days with heavy rain various activities would be laid on, and students would rotate round these (Fig. 2.3). The sorting of samples linked to the programme of flotation and wet screening was a routine activity, but a significant number of samples were always held back, at least until towards the end of the season, so that they would be available for wet-weather days. The same applied to artefact processing, and data entry onto laptop computers, though this was necessarily only open to few students at any one time. Field school students were also able to read selected texts relevant to their written assignments, and short impromptu seminars on aspects of the cultural history of the region or particular methods could be held. A small collection of animal bone and ceramics of other periods and regions than that of the site was also kept at the base camp, so that training on identification of these could take place. In addition, a relatively simple and a more complex profile drawing was available in multiple copies so that students could be taught the principles of the Harris matrix, and then could apply this to the exercises; if this had already been part of their training, the more complex example could challenge their understanding and application of the method. If the rain eased off but the site was still too waterlogged, visits to other sites or the teaching and learning of survey techniques could also take place. Many of these activities required some planning, but were well worth having ready even though some years they were not required at all. These types of activity may be necessary if transport from base camp to site breaks down, or if an activity such as survey has to be halted because of equipment failure or staff illness. Just having some back-up plans, for which sufficient staff are proficient, creates a sense of confidence and allows the resolution of the problem to be the focus of the director, rather than pacifying students and keeping up their morale (see also Chap. 11).

Effective teaching and learning is always a fluid, adaptable exercise, though linked back to core aims and objectives. The very definition of those objectives needs to be informed by risk awareness, and back-up plans also need to be designed to replace or augment aspects that may for unforeseen reasons be unavailable. The writing of the programme outlines needs to factor in the possibilities of flexible delivery so that the primary goals can be satisfied, and that only a number, though not necessarily all practical aspects of the syllabus, may be provided to fulfil the learning objectives. Students also must appreciate that understanding this flexibility and adaptation to conditions is part and parcel of a field school experience, and that while staff attempt to provide every aspect of training, this is not always within their power. Communication on all sides is essential to ensure a positive attitude in the face of adversity.

2.6   Conclusions

The field school is a highly effective learning environment in more ways than most students or staff appreciate. Making explicit to students all the ways in which they learn about archaeological methods, themselves, other people, and the context within which fieldwork takes place is important. They may even learn about the past, though that may be as effectively taught in the classroom. Field school directors should consider how they frame the aims and objectives of their programmes and what literature about the project and the training should be provided ahead of being in the field (which will be read avidly by some and not at all by others). Once in action, the staff should explain how the work being undertaken by the students is an ongoing and deepening learning experience, despite its sometimes repetitive character. Modes of assessment should be linked to the aims and objectives and be clear to students and all staff so that levels of attainment are fair for all. The variety inherent in a field school means that many can achieve proficiency in at least some aspects of the programme, and it is an arena in which some who find class-based working difficult can flourish. Indeed, the field school may be a better indicator of general employability than some classroom contexts, as practical problem-solving skills and the ability to work in a team and meet objectives within time constraints are highly valued (see Chap. 3). Thus, at the level of formal and informal teaching and learning, and as an assessor of employability, the field school can be extremely instructive for all.

References

Adler, M. (2001). The register. Certifying your archaeology field school. *SAA Record, 1*(2), 15–16.

Baxter, J. E. (2009). *Archaeological field schools. A guide for teaching in the field*. Walnut Creek: Left Coast Press.

Boyle, A., Maguire, S., Martin, A., Milsom, C., Nash, R., Rawlinson, S., Turner, A., Wurthmann, S., & Conchie, S. (2007). Fieldwork is good: The student perception and the affective domain. *Journal of Geography in Higher Education, 31*(2), 299–317.

Cobb, H., & Richardson, P. (2008). Transition/transformation: Exploring alternative excavation practices to transform student learning and development in the field. *Research in Archaeological Education, 2*, 21–40. Retrieved March 21, 2011, from http://www.heacademy.ac.uk/hca/archaeology/RAEJournal

Darvill, T. (2008). UK Archaeology benchmark updated. *Research in Archaeological Education, 1*. Retrieved March 21, 2011, from http://www.heacademy.ac.uk/hca/archaeology/RAEJournal

Fleming, N., & Baume, D. (2006). Styles again: VARKing up the right tree! *Educational Developments, 7*(4), 4–7.

Fuller, I., Book, M., & Holt, K. (2010). Linking teaching and research in undergraduate physical geography papers: the role of fieldwork. *New Zealand Geographer, 66*, 196–202.

Hamilakis, Y. (2004). Archaeology and the politics of pedagogy. *World Archaeology, 36*(2), 287–309.

Healey, M. (2005). Linking teaching and research to benefit student learning. *Journal of Geography in Higher Education, 29*, 183–201.

Healey, M., & Jenkins, A. (2000). Kolb's experiential learning theory and its application in geography in higher education. *Journal of Geography, 99*, 185–195.

Honey, P., & Mumford, A. (1982). *Manual of learning styles*. London: P. Honey.

Kolb, D. A. (1984). *Experiential learning: Experience as the source of learning and development*. New Jersey: Prentice-Hall.

Marvell, A. (2008). Student-led presentations in situ: the challenges to presenting on the edge of a volcano. *Journal of Geography in Higher Education, 32*(2), 321–335.

Mills, B. J., Altaha, M., Welch, J. R., & Ferguson, T. J. (2008). Field schools without trowels: teaching archaeological ethics and heritage preservation in a collaborative context. In S. W. Silliman (Ed.), *Collaboration at the trowel's edge: Teaching and learning in indigenous archaeology* (pp. 25–249). American studies in archaeology, J. Ware, general editor. Tucson: University of Arizona Press.

Neusius, S. W. (2009). Changing the curriculum: Preparing archaeologists for careers in applied archaeology. *SAA Record, 9*(1), 18–22.

Perry, J. E. (2004). Authentic learning in field schools: Preparing future members of the archaeological community. *World Archaeology, 36*(2), 236–260.

Prosser, M. (2010). *Grade descriptors and standards*. University of Hong Kong, Hong Kong. Retrieved April 24, 2011, from http://www.cetl.hku.hk/system/files/Grade_Descriptors_and_Standards_Oct_2010.pdf

Schuldenrein, J., & Altschul, J. (2000). Archaeological education and private sector employment. In S. J. Bender & G. S. Smith (Eds.), *Teaching archaeology in the twenty-first century* (pp. 59–64). Washington: Society for American Archaeology.

Silliman, S. W. (Ed.). (2008). *Collaborating at the trowel's edge: Teaching and learning in indigenous archaeology*. University of Arizona Press, Tucson.

Spronken-Smith, R., & Hilton, M. (2009). Recapturing quality field experiences and strengthening teaching-research links. *New Zealand Geographer, 65*, 139–146.

Walker, M., & Saitta, D. J. (2002). Teaching the craft of archaeology: Theory, practice, and the field school. *International Journal of Historical Archaeology, 6*(3), 199–207.

Chapter 3
Field Schools, Transferable Skills and Enhancing Employability

Hannah Cobb and Karina Croucher

3.1 Introduction

Recent archaeological literature in the UK has begun to draw attention to the fact that despite the recognised importance of fieldwork, little research has been undertaken into fieldwork processes and experiences. For example, in the volume *Critical Approaches to Fieldwork,* Gavin Lucas discusses how despite theorising archaeological interpretation, little is done to really analyse or examine how fieldwork is undertaken today (Lucas 2001:1–2). In the last decade, there have been some developments in this area which build on earlier observations by Hodder in 1997 (see Andrews et al. 2000; Bender et al. 2007; Cobb and Richardson 2009; Cobb et al. in press Lewis 2006 for examples of some accounts that have tried to address this issue), yet even as the discipline moves towards a more explicit approach to theorizing field practice, archaeological field training and the role of fieldwork in degree programs have received little consideration.

Despite the lack of explicit theorization, the Subject Benchmarking Statement (QAA 2007, and see Darvill 2008 for a summary of recent updates to the Statement), produced in the UK by the government, recognises the important role that fieldwork plays in the undergraduate degree. The statement asserts that

> … much of the best teaching and learning in archaeology will be an interactive process from which students and academics gain mutual benefit because of the research led environment for teaching. Students need to be encouraged to learn through experience, both as individuals and as members of defined teams, with practicals and fieldwork playing important roles in such provision (QAA 2007).

H. Cobb • K. Croucher (✉)
Archaeology, School of Arts, Histories and Cultures, University of Manchester, Oxford Road, Manchester M13 9PL, UK
e-mail: Karina.croucher@manchester.ac.uk

H. Mytum (ed.), *Global Perspectives on Archaeological Field Schools: Constructions of Knowledge and Experience*, DOI 10.1007/978-1-4614-0433-0_3, © Springer Science+Business Media, LLC 2012

There have been valuable local studies undertaken (Brookes 2008; Thorpe 2004), which support such assertions. However, even with the QAA supporting the importance of practical work, there was, until 2004, a lack of any real data on undergraduate fieldwork experiences at a national level in the UK.

This is certainly problematic given that a practical fieldwork training element is a central component of most single and joint honours archaeology degrees in the UK. Here undergraduate degrees are usually 3 years in length. They will typically include a narrower breadth of subject coverage than North American degrees, for instance, but will almost always include a taught component in archaeological field skills, or more generally vocational skills training. This can be provided in different ways, ranging from entire modules spent in the field during the academic year, to the more common format of teaching a field skills module in the classroom during the academic year, which is then complemented by (and often assessed during) a compulsory element of field training under taken during the summer months. The British academic year certainly lends itself to this well, given that it begins in September and ends in June, thus providing a long summer period during which training excavations are normally run. Of course the length of the summer is not necessarily a reflection of the length of fieldwork students are required to undertake, and this can vary quite extremely from 2 to 12 weeks over the entire length of the degree. In some degree programmes, students are encouraged to take even longer in the field, sometimes up to a year long placement in industry to develop their field kill.

Given the centrality of fieldwork in the disciplinary culture of archaeology, its role in the undergraduate degree, and the education vs. training debate in British archaeology (Aitchison 2004; Hamilakis 2004; Hamilakis and Rainbird 2004:52; Dowson et al. 2004; Stone 2004:6; Rainbird and Hamilakis and references within 2001; Collis 2000), examining what students actually want from their degrees is vitally important. Consequently, during the summer months of 2004 and 2005, the archaeology team in the History, Classics and Archaeology Subject Centre of the Higher Education Academy (HEA) carried out the most comprehensive survey to date of the opinions and experiences of archaeological fieldwork among archaeology students and staff in the UK (Croucher, Cobb and Brennan 2008). Our aim was to investigate perceptions and expectations of fieldwork in archaeology at undergraduate degree level in Britain. To do this, we visited 32 excavations that were either explicitly run as field schools or that provided training opportunities for archaeology undergraduates.

As well as being driven by the needs of archaeology departments and students, this project also arose out of a growing concern from archaeological employers that the graduates they are employing are felt to be inadequately equipped for a career in archaeology (Aitchison 2004, 2008). Consequently, by investigating the role of fieldwork and vocational training, the project aimed to develop a greater understanding of the debate, considering the positions, responsibilities and restrictions on universities, as well as the perspectives of students and staff on the issue of vocational training. This chapter highlights some of our findings, with a particular focus on transferable skills and employability.

3.2 Investigating the Role of Fieldwork in Teaching and Learning Archaeology: Methodology

To undertake a comprehensive survey that addressed staff and student expectations of the fieldwork experience, we decided we had to *speak* to both staff and students, rather than simply circulating questionnaires and/or reading course handouts. Interviewing people face-to-face would allow them to be more relaxed and forth-coming in their responses "in conversation" rather than having to find the time to write down their responses on paper.

Once we had decided that we needed to speak directly to staff and students, the location was considered; should we simply speak to people while at university? Although we are aware that speaking with students and staff in the university environment does have its merits, mainly in offering a distanced perspective, for this particular study we felt that gaining immediate responses was preferable. Consequently, we felt that through interviewing in the field, students would not feel the same restraint placed on them as by a classroom location. It is all too easy to gain a distorted picture of fieldwork once back at university, and while memories of the highs and lows may last, details of individuals' thoughts, opinions and experiences in the field soon fade. We therefore felt that speaking to staff and students while actually on site would allow us direct access to actual experiences. Following this decision, we advertised the project to all Higher Education Institutions (HEIs) offering archaeology in the UK and then responded to invitations from project directors to attend their excavations.

It quickly became clear that attending the excavations in person also gave us the invaluable opportunity to observe more subjective data – the general feeling of the site, attitudes, and emotions – essential components of any dig that could be lost in questionnaires. We participated in the projects as observers, and as we are all archaeologists, could situate ourselves within the site or lab dynamic. While our very being on site would have had some influence, it is hoped that our relaxed and informal approaches, and experiences of fieldwork, would enable greater acceptance and thus access to the opinions and experiences of students and staff. We could therefore pick up on the mood or "vibe" of the site, observe how students interacted with staff, as well as ask more detailed questions if we thought it was appropriate (see Edgeworth 2003, 2006; Everill 2006; Holtorf 2006 in the use of participant observation in relation to research into professional archaeology).

HEA staff undertook site visits over the summers of 2004 and 2005, visiting a total of 32 sites, and speaking with 434 students and 103 staff, representing 25 UK HEIs, 9 Further Education (FE) and Continuing Education (CE) institutions, 4 non-UK HEIs and 4 non-student volunteers (for further information on the demographics of participants see Croucher et al. 2008: Figs. 3.2–3.4). Of the students interviewed, 202 of these were entering their second year of study, and 175 their third year. These projects represent a broad spectrum of fieldwork approaches, all demonstrating different methods of training, with a wide range of tasks undertaken by students, including trowelling and excavation, surveying, planning and drawing, running visitor tours, and for some, website updates. The running of these projects also covered a range of approaches; from the pure research project to the "summer school" directly set up to train students in archaeological techniques.

The questionnaires covered a variety of topics, from basic demographic questions to more in-depth interrogations of what was expected from fieldwork. Questions addressed whether fieldwork should be compulsory, the assessment of fieldwork, the length and amount of fieldwork, the role of fieldwork with relation to archaeological and non-archaeological careers, issues of responsibility, the role of professional contract archaeology organisations, feedback, likes and dislikes of the fieldwork experience, integration of fieldwork into the rest of the course, the implications of fees, and student opinions of their contribution to the bigger archaeological picture.

The process through which students and staff were selected for interview was largely random. At each site, we aimed to interview at least one third of all students present and as many staff as possible. However, in general we adopted a flexible attitude toward questioning staff and students; sometimes questioning participants as they dug, sometimes questioning them during break times and sometimes taking them aside while digging was going on. We subsequently evaluated the material and responses gathered to assess trends and perspectives, rather than focusing on individual institutions or projects. Our aim was not to "name and shame" departments where students highlighted negative experiences (as inevitably some did). The very involvement of sites and departments in this project, enabling us to interview and participate on site, demonstrates the commitment of all departments involved to providing a positive fieldwork experience for their students. Instead it should be noted that the negative responses we did receive (which were in a minority) provided as much valuable evidence as those cases of good practice, and these formed an essential component in informing our recommendations.

3.3   Investigating the Role of Fieldwork in Teaching and Learning Archaeology: Key Findings

This project has been the most wide-ranging exploration of archaeological staff and students in the UK. Based on figures for 2004/2005 from the Higher Education Statistics Agency, we interviewed over 10% of full-time UK archaeology undergraduates during the survey phase. This provided us with the opportunity to investigate the current state of practical provision and analyse the experiences of staff and students throughout the country. Some of our findings relate specifically to elements of the British Higher Education system (e.g. demographics, the role of tuition fees, etc.) and as the remit of this volume is for an international audience, these findings will not be discussed here (see Croucher et al. 2008 for further details). However, some of our findings are clearly applicable to the broader training of archaeology students, wherever their archaeological field school is held. In particular, we have identified issues of employability and transferable skills. Two specific points form the basis of our findings: fieldwork training has a significant impact on student career choices; and despite this, students have trouble identifying that fieldwork training provides a high level of skills transferable to non-archaeological career paths. In the rest of this section, we will outline key findings on these.

3.3.1 Fieldwork Training, Archaeological Careers and Employability

Fieldwork has a huge role to play for students in influencing whether they wish to pursue an archaeological career. Our results showed that 58% felt that fieldwork has a positive influence on their decision to consider an archaeological (or related) career, with 29% finding their fieldwork had a negative impact on their choice to pursue archaeology. Just 13% felt that fieldwork did not have an impact on their career choices. These results are mirrored by research undertaken by Jackson into archaeology graduates. Of the 710 interviewees who had graduated in an archaeology or related subject, 92.5% had undertaken fieldwork, with 63% citing their experience as influencing their career choices (Jackson and Sinclair 2009:12). It is clear that fieldwork itself plays an important role in student career decisions, whether to pursue archaeology, or to consider a different area of employment.

The model of academic departments working with professional units is an ideal situation. Through involving archaeological employers in training students, universities can benefit from a wider skills-base and the employers can help train the archaeological workers of the future. Students gain a greater range of skills and techniques, as well as contacts and career guidance (of both the positive and negative aspects of a career in the field). However, it is ideal when university staff are also involved, with a good balance between understanding the practical aspects involved in fieldwork alongside the larger research frameworks being investigated. Additionally, for those students who do wish to gain extra fieldwork experience, it is profitable for universities to have greater links with both the profession and with other universities, with exchanges of students and skilled staff across excavations. This also provides an avenue for those wanting more specialised training, enabling easier access to a wider range of experiences that would allow them to make informed career choices.

Although at present there are no precise figures available, current estimates over the last decade have suggested that consistently only approximately 15% of archaeology graduates tend to follow a career in archaeology (Collis 2001). However, the actual figure may be higher, as suggested by Jackson and Sinclair (2009), where 39% of respondents were in archaeological careers, and a further 11% were potentially archaeology-related, although the sample may show a bias towards those remaining in contact with the archaeology sector (Jackson and Sinclair 2009), suggesting that at best only 50% of archaeology graduates remain in a related field. Our study demonstrated that while in the field student career aspirations are, at least temporarily, more focused towards an archaeological career path, in particular, as Fig. 3.1 demonstrates, 57% of the 434 students interviewed stated that they intended

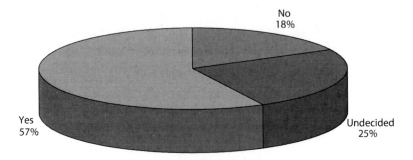

Fig. 3.1 Student responses to the question: "Do you wish to follow a career in archaeology?"

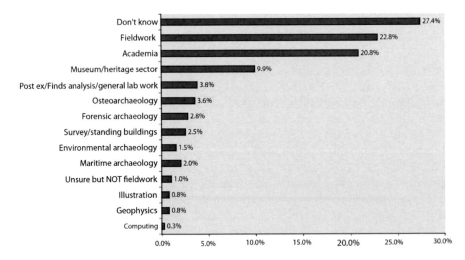

Fig. 3.2 Breakdown of the various archaeology sectors that students want to work in

to continue with a career in the subject, with a further 25% who were undecided. Significantly, only 18% of students offered a definitive "no" to following an archaeological career. Of those who intended to, or thought they might follow a career in archaeology, just over a quarter (27%) were unsure as to what area of the discipline they wanted to follow, however just over a fifth of all of those wanting to work in archaeology wished to follow a career in fieldwork (22.8%), and a similar number (20.8%) wanted to follow an academic route. The remaining 30% of students expressed interests in careers in the museum/heritage sector, and additionally a pursuit of specialisms, with finds-based options being particularly popular (Fig. 3.2). Research carried out by Jackson and Sinclair (2009) into archaeological graduates revealed that of 710 interviewed, 50% had wanted to become archaeologists at the beginning of their degrees, a figure rising to 55% by the completion of their degrees. Those not wanting a career involving archaeology rose from 16% at the start of their degrees to 30% by graduation (Jackson and Sinclair 2009:11).

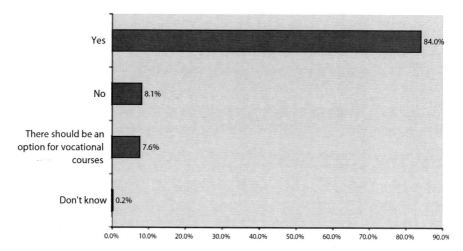

Fig. 3.3 Student responses to the question: "Should universities prepare students for a career in archaeology?"

While these results indicate that fieldwork training has a significant impact on student career choices, this raises a crucial dilemma faced by universities globally in terms of their role in developing employability in their degree programmes. While only a small percentage pursue archaeological careers (as low as 15%), the archaeological sector is still likely to be the largest single area of employment. This situation raises questions regarding the responsibilities of universities, and the archaeological profession, in terms of training. To examine this, we asked staff and students whether a degree should prepare students for a career in professional archaeology. As Fig. 3.3 demonstrates, an overwhelming 84% of students felt that it was the responsibility of the university to prepare them for an archaeological career. Here students regularly stated that "you are doing an archaeology degree so [it] should prepare you for a career in it" (AB501), and that "if I wanted a less vocational course I would have done something else" (JW511).

Student views contrast significantly with staff opinions (Fig. 3.4). Only 36% of staff felt that a degree in archaeology actually does prepare a student for an archaeological career, and 19% suggested that a degree only sometimes (depending on the student and/ or institution) prepares the student for a career in archaeology. For those 18% who suggested an archaeology degree provided students only with "the basics", many suggested that this was because vocational training was an ongoing process. Here, staff such as AB198L argued that in undergraduate training "we go some way – producing apprentices, not excavating archaeologists". Although for the 26% who felt that archaeology did not prepare students for a career in archaeology, many argued that "it shouldn't". Staff members cited reasons such as "few other degrees produce practicing professionals; a degree is a foundation for the career. MA courses could prepare better" (KC019V). What seems most critical here is the clear disparity that exists in staff and student expectations as to the role of fieldwork within the undergraduate degree. Moreover, it is clear that there is little unity among staff in general as to the role the undergraduate degree should play in preparing students for a career in archaeology.

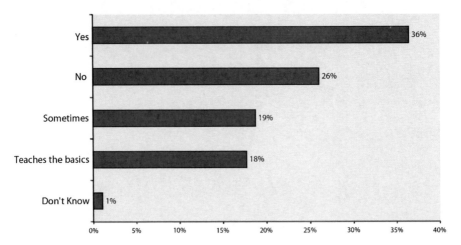

Fig. 3.4 Staff responses to the question: "Does a degree prepare students for a career in archaeology?"

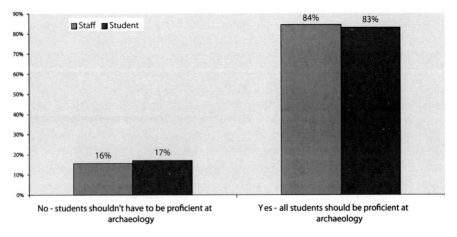

Fig. 3.5 Student responses to the question: "Should students be proficient in practical aspects of archaeology on completing their degree?"

This lack of consensus in the role of the undergraduate degree and fieldwork training more specifically is clearly troubling. Nonetheless, staff and students did agree when it came to considering the level of fieldwork proficiency students should obtain upon graduating. While many students recognised they would not be undertaking a career in archaeology, 83% still felt that having an archaeology degree should mean that students leave university being proficient in archaeological practices. This was also reflected in staff attitudes, with 84% believing students should be proficient at fieldwork when finishing their degree (Fig. 3.5). Overwhelmingly then, both staff and students (regardless of whether or not students wanted a career in archaeology) felt that archaeology graduates should be proficient in fieldwork when leaving university. This stands in stark contrast to the previous statistic that showed 36% of staff thought that a degree did not prepare students for a career in archaeology, and in turn returns us to the key dilemma; should there be a global consensus on the responsibility of universities, and the archaeological profession, in terms of training undergraduates for a possible career in archaeology? Perhaps the most straightforward response to this question arises from our study; here we found that staff and students alike are confident of the importance of undertaking practical training to a reasonable level. Thus, while the definition of this level varies between students, universities and employers, striving for proficiency in core field skills provides at least some answer to the basic level of responsibility universities should have in preparing students for an archaeological career.

3.3.2   Fieldwork Training, Non-archaeological Careers and Transferable Skills

As we have discussed above, while a large number of students wish to follow a career in archaeology upon graduation, a great percentage of students do not continue into professional archaeology. Consequently, we examined the key question of the role fieldwork plays in equipping students with the generic and transferable skills that will be important no matter which career they choose.

Research carried out in 2007, surveying 710 graduates who had obtained an archaeology degree since 2000, revealed that those who didn't enter archaeology or a related field were employed in a range of sectors including business, marketing and finance, health and social care, law, IT and leisure and tourism (Jackson and Sinclair 2009:27). Both Jackson and Sinclair (2009:24) and Croucher et al. (2008) demonstrate that the skills archaeology graduates obtain are relevant to other careers. In a study of entrepreneurial employment routes in the humanities, fieldwork was repeatedly cited by many graduates as developing transferable skills (Croucher et al. 2008:17). However, there are certain steps that can be taken to ensure the most is gained from fieldwork for the future employability of students. These include assessment, reflexivity and communication.

In an increasingly competitive graduate employment market, an awareness of the transferable skills that an undergraduate degree provides significantly enhances students' employment chances. An archaeology undergraduate degree, and the practical component of this in particular, can provide a wide range of transferable skills that can be applied within other career paths (Aitchison and Giles 2006). Our study sought to examine whether students were aware of this, and how they felt their degree may enhance their employability. We explicitly asked what transferable skills were being acquired during fieldwork. When student and staff responses are compared on this question, the results mirror one another, with both staff and students citing teamwork most frequently. Following this, most students saw that they were gaining archaeological skills and general communication and social skills. Although less frequently cited, between 8 and 5% of student responses also noted aspects such as analysis, observation, initiative, organisation and responsibility, as key transferable skills that fieldwork provided them with (Table 3.1). More significant

Table 3.1 Responses by staff and students to the question "What transferable skills does fieldwork provide?" The relevant QAA Archaeology benchmark statements are given in italics

	Staff%	Student%
Teamwork	24.4	25.5
Collaborate effectively in a team via experience of working in a group, for example through fieldwork, laboratory and/or project work		
Social/communication skills	17.6	10.0
Present effective presentations for different kinds of audiences; (as fieldwork often involves working in new environments with minimal support) appreciate and be sensitive to different cultures, and deal with unfamiliar situations		
Employability in archaeology/fundamental archaeological skills	7.4	10.7
Observation/analysis/recording skills	11.9	7.3
Practice core fieldwork techniques of identification, surveying, recording, excavation, and sampling; practice core laboratory techniques of recording, measurement, analysis, and interpretation of archaeological material; observe and describe different classes of primary archaeological data, and objectively record their characteristics		
Independence/confidence building/initiative	6.3	4.9
Physical/hard work/hands-on skills	0.0	5.9
Organization/multi-tasking	4.0	4.7
Responsibility/leadership/management skills	5.1	4.4
Ability to work under pressure/persevere in hard conditions/commitment/ determination	0.0	4.6
Learn to take orders/work in a disciplined environment	2.3	3.9
Time management	2.3	2.8
Problem solving	2.8	1.8
Draw down and apply appropriate scholarly, theoretical, and scientific principles and concepts to archaeological problems		
Numeracy skills	2.3	1.9
Select and apply appropriate statistical and numerical techniques to process archaeological data, recognizing the potential and limitations of such techniques		
Patience/accuracy	0.0	2.0
General (not listed) transferable skills	0.0	1.7
Written skills	0.0	1.2
Prepare effective written communications for different readerships		
Health and safety	1.7	0.8
Appreciate the importance of safety procedures and responsibilities (both personal and with regard to others) in the field and the laboratory		
Skills relating to other professions	0.0	1.1
Life skills/personal development	0.0	1.0
No skills	0.0	0.9
Computing skills	0.6	0.8
Make effective and appropriate use of C&IT (such as word processing packages, databases, and spreadsheets)		
Wider understanding of subject	2.8	0.4

(continued)

Table 3.1 (continued)

Surveying skills	4.5	0.0
Practice core fieldwork techniques of identification, surveying, recording, excavation, and sampling		
Research skills	1.7	0.4
Assemble coherent research/project designs		
Don't know	0.0	0.6
Only transferable skills relevant to (specific area of) archaeology	0.0	0.6
photography	1.1	0.0
Make effective and appropriate forms of visual presentation (graphics, photographs, spreadsheets)		
Finances	0.6	0.0
Assemble coherent research/project designs		
Interpretation skills	0.6	0.0
Discover and recognize the archaeological significance of material remains and landscapes; interpret spatial data, integrating theoretical models, traces surviving in present-day landscapes, and excavation data		

are the skills that few students mentioned. Less than 2% felt that fieldwork provided life skills, written, research, and numeracy skills, for instance. Also of concern are the gaining of abilities such as critical thinking, independent thought, and problem solving, with many students not realising, or at least not articulating, the role that fieldwork may play in developing these. Even more concerning was the small percentage of respondents that said they did not know what transferable skills fieldwork provided them with or that it did not provide any transferable skills at all. Student AB141, for example, said "you don't pick up many transferable skills in fieldwork – unless you want to be a navvy".

That students have little recognition of transferable skills is clearly problematic. In a global climate of financial downturn and recession, being aware of and then able to develop and maximise the skills fieldwork provides will ultimately be an important factor in enhancing student employability, whether students want to follow an archaeological or non-archaeological career path. Perhaps then in answer to the question posed above of whether there should be a global consensus on the responsibility of universities in terms of training undergraduates, a key responsibility could be to foster a more explicit understanding of the transferability of skills learnt in fieldwork (Table 3.1). In addition, we identified a series of other areas of fieldwork training that can be enhanced to develop student employability and awareness of the transferability of skills learnt in fieldwork. These include considering the roles of assessment, reflexivity, and communication in enhancing undergraduate understandings of the varied skills that fieldwork can provide.

Through stimulating reflexivity in the learning process, students are encouraged to consider their learning accomplishments. This includes recognising how they learn, and assessing their achievements. It is often the case that students are not aware of the transferable skills they are gaining through their fieldwork experiences (Table 3.1) or view their skills in very simplistic terms. For instance, while they may

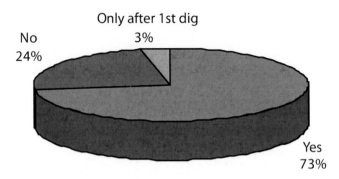

Fig. 3.6 Student responses to the question: "Should fieldwork be assessed?"

be aware that they can survey a grid or sort finds, they may not relate this to planning and organisation, analysis, and confident independent working or team work. Yet these are the broader skills that employers require interviewees to articulate. Through encouraging a reflexive approach, students are more likely to be able to recognise and articulate the skills they are gaining, as well as those that may still be needed. A key method for encouraging reflexivity is through assessment.

Assessment can play a key role in encouraging students to reflect on their fieldwork experiences. During our survey, we asked about the role of assessment, with 49% of students being assessed on their fieldwork, and 45% not being assessed (those that were not assessed also included second- and third-year students who had been assessed previously, but were not being assessed for their current excavation). Alarmingly, 6% of students did not know whether they were being assessed or not, which highlights communication issues within some institutions. We also asked students whether they felt their fieldwork *should* be assessed. Overwhelmingly, 73% answered that it should be (Fig. 3.6), with comments made including: "fieldwork should be assessed so you can see how much you have learnt" (student AB041) and "assessment is a reward for all of your effort" (student AB122). It was felt that progression could be both demonstrated and realised through assessment, as well as identifying areas for improvement.

During our study, it became apparent that assessment played an important role in motivating students, especially if fieldwork was taking place during vacation time and was compulsory. However, if it had no bearing on the outcome of their course, then students often seemed to lose interest very quickly. Incidences of resentment and anger at being "made" to undertake practical work were not uncommon. We also asked students "how does fieldwork relate to the rest of your course?", and "what are you contributing to the bigger picture?" Those answering negatively to these questions were repeatedly those students who were not being assessed. The relationship is not clear-cut and student experiences are not solely dependant on assessment, yet when assessment is in place, students are generally more positive and have a better understanding of their role within the archaeological project. This is also related to issues of communication, where students need to understand the project as a whole, and their personal contribution to it.

The assessment of fieldwork can also be especially beneficial for those who learn in different ways. Fieldwork provides real, hands-on experiences, making learning more tangible, especially valuable for visual and experiential learners; such experience is an essential component in the learning cycle of many students (Boud et al. 1985; Honey and Mumford 1982; Kolb 1984). As well as developing social, personal and communicative skills, fieldwork also demonstrates the link between theory and practice, helping students to understand field reports, and crucially promotes understanding of the methods and processes behind the creation of archaeological knowledge (Chap. 2). This enables students to develop critical thinking, analytical and interpretative skills and abilities, as well as providing an important arena for students to develop their understanding of archaeological career paths.

Through integrating practical training into the degree as something that can be graded, the students were given an opportunity to prove themselves in a forum other than a classroom (see Thorpe 2004 for a discussion of methods used to grade practical work). From our interviews, we heard comments from students who were not necessarily good at essay writing or more traditional academic pursuits, but proved themselves to be excellent students when given a practical task. By assessing or grading practical performance, it gives the students another chance to excel, using a set of skills that may not be developed through classroom learning. These experiences can be crucial for later employment and should be encouraged, giving students opportunities to excel in a wider field of activities.

Assessment can also be seen as a mechanism for ensuring that all students are aware of their roles on the project, and what they are learning. While these should happen regardless, in reality, students can often feel unguided; assessment provides an additional framework that ensures students are encouraged to think about their aims and achievements. If students are aware and thinking about their skills and achievements, including the transferable skills they are obtaining, they are already a step closer to being able to articulate these to future employers, or indeed to recognise the wide variety of other types of employment open to archaeology graduates.

There are various ways that assessment is carried out. Some excavation projects used log books or passports to assess tasks undertaken directly on site. Others used reflective journals alongside these, encouraging students to think about what they had learnt and areas for improvement. It is also good practice to relate assessment outside of the field to fieldwork, with a closer integration between fieldwork and the rest of the degree programme. It is important that students do not feel that their fieldwork is isolated and unrelated to the rest of their degree. Ideally, the relationship between fieldwork and the rest of the degree programme should be clear, with students able to see the connections between field and class work. Examples of good practice were seen when there were clear and explicit links between the fieldwork being undertaken and the rest of the degree programme. Negative experiences often revolved around a lack of understanding of the relevancy of their fieldwork, often closely, but not exclusively, linked to the issue of assessment.

Communication is also a vital element that should be further developed through the fieldwork experience. As well as encouraging reflection and an understanding of the skills gained, communicating these, both through writing and verbally, are essential skills for graduates. Assessment often plays some role, especially with written communication (although presentations should also be encouraged), but there are other ways that students can be encouraged to reflect and communicate on site. Involving students in guiding visitors can have a huge impact, not just in their communication skills, but in encouraging students to see the bigger project, and their individual roles within it. Including students in site tours is also a positive step, encouraging students to explain their area of the trench to the rest of the group, including debates about ambiguities that rise, with students thinking about the interpretative process that happens both on and off site. Related to this point is the importance of giving students responsibility during the excavation process. A significant source of resentment by students arose from being removed from the trench once anything "interesting" or "important" was discovered. On occasion this is inevitable, as some things are too rare for less-than-expert attention. However, in the majority of cases, it would be feasible for the student to continue with excavation under supervision, or indeed work alongside the expert, thus allowing them to learn, and to see the process of excavation through. Additionally, crediting students with the role they have played is has a positive impact, for instance, including their names in site reports, as seen at some of the leading sites. Students who realised their names would be in print generally took a much more active and responsible role towards the excavation.

3.4   Conclusion

Through our survey of students and staff on archaeological excavations, there is no room for doubt as to the unique value and importance of the role of fieldwork in the archaeological degree. As well as providing social and personal development (an area that is a huge strength of archaeology as opposed to other subjects studying the past), fieldwork offers real professional development. Our survey illustrates the extent to which fieldwork provides both vocational experience and transferable skills. It is also fundamental in encouraging an understanding of the production of knowledge in the discipline. Comprehending the nature of archaeological excavation, the role of interpretation, and the idea that not everything is always factual or clear-cut is central to academic research in archaeology, and this is something the students stated they only fully comprehended after being in the field.

However, fieldwork experiences can usually be improved. Research shows that reflexive learning can develop the ability of students to recognise and build on the skills they are gaining (Kolb 1984; Honey and Mumford 1982). Assessment can play a key role, as it can encourage students to communicate and articulate the skills they have gained. The ability to recognise and communicate the vocational and transferable skills gained through fieldwork and the archaeological degree is essential for students graduating today, especially pertinent in the current global economic climate, where competition for jobs is set to become fiercer. It is essential that the unique skills offered through excavation are maximised and, crucially, recognised and communicated by those embarking on their new careers, either as archaeologists or in the diverse range of other career paths available.

Acknowledgements First and foremost we thank the numerous staff and students who so willingly participated in this research project and made us so welcome on their excavation sites. The project began under the direction of Thomas Dowson as the Archaeology Subject Director within the Higher Education Academy's Subject Centre for History, Classics and Archaeology. It was managed by Karina Croucher and undertaken by Ange Brennan, Karina Croucher and Hannah Cobb. We would also like to thank Sarah Croucher, Paul Goodwin, Chris Jones, Helen Lee, Steven Price and Joanna Wright who carried out additional interviews, and Lucy Day for typesetting the illustrations that are used in this paper and that appeared in the original report. The project was inspired by comparable research into fieldwork by the Geography, Earth and Environmental Sciences Subject Centre and by our colleagues in the History section of the Subject Centre who had carried out a survey of staff and student expectations of the value of lectures.

References

Aitchison, K. (2004). Supply, demand and a failure of understanding: Addressing the culture clash between archaeologists' expectations for training and employment in 'academia' versus 'practice'. *World Archaeology, 36*(2), 219–230.

Aitchison, K., & Giles, M. (2006). Employability and curriculum design. *Guides for teaching and learning in archaeology number 4.* Higher Education Academy. Retrieved April 21, 2011, from www.heacademy.ac.uk/hca/archaeology/features_resources/guides.

Aitchison, K. (2008). Workforce issues in archaeology and historic building conservation. Retrieved 21, April 2011, from http://www.creative-choices.co.uk/server.php?show=ConWebDoc.490.

Andrews, G., Barrett, J. C., & Lewis, J. S. C. (2000). Interpretation not record: The practice of archaeology. *Antiquity, 74*, 525–530.

Bender, B., Hamilton, S., & Tilley, C. (2007). *Stone worlds: Narrative and reflexivity in landscape archaeology.* London: Berg.

Boud, D., Keogh, R., & Walker, D. (Eds.). (1985). *Reflection: Turning experience into learning.* London: Kogan Page.

Brookes, S. (2008). Archaeology in the field: Enhancing the role of fieldwork training and teaching. *Research in archaeological education* (Vol. 1). Retrieved April 21, 2011, from http://www.heacademy.ac.uk/hca/archaeology/RAEJournal.

Cobb, H. L., & Richardson, P. (2009). Transition/transformation: Exploring alternative excavation practices to transform student learning and development in the field. *Research in archaeological education* (Vol. 1). (2) Prehistoric pedagogies? Approaches to teaching European prehistoric archaeology, pp. 21–40. Retrieved April 21, 2011, from http://www.heacademy.ac.uk/hca/archaeology/RAEJournal.

Cobb, H. L., Harris, O., Jones, C., & Richardson, P. (in press). *Reconsidering fieldwork: Exploring on site relationships between theory and practice.* New York: Springer.

Collis, J. (2000). Towards a national training scheme. *Antiquity, 74*, 208–214.

Collis, J. (2001). Teaching archaeology in British universities: A personal polemic. In P. Rainbird & Y. Hamilakis (Eds.), *Interrogating pedagogies: Archaeology in higher education.* British Archaeological Reports International Series 948. (pp.15–20). Archaeopress, Oxford.

Croucher, K., Canning, J., Gawthrope, J., Allen, R., Croucher, S., & Ross, C. (2008). Here be dragons? Enterprising graduates in the humanities. Southampton; subject centre of languages, linguistics and area studies. Retrieved April 21, 2011, from http://www.llas.ac.uk/publications/enterprise.html.

Darvill, T. (2008). UK Archaeology benchmark updated. *Research in archaeological education*, vol. 1. Retrieved April 24, 2011, from http://www.heacademy.ac.uk/hca/archaeology/RAEJournal.

Dowson, T., Johnson, M., Aitchison, K., Walker, J., Hanson, W., Benjamin, R., et al. (2004). *Creating tomorrows archaeologists: Who sets the agenda?* Retrieved April 24, 2011, from http://www.heacademy.ac.uk/hca/events/detail/past/creating_tomorrows_archaeologists_18_12_2004.

Edgeworth, M. (2003). *Acts of discovery: An ethnography of archaeological practice*. Oxford: Archaeopress.

Edgeworth, M. (2006). *Ethnographies of archaeology practice: Cultural encounters, material transformations*. New York: Altamira.

Everill, P. (2006). *The invisible diggers: Contemporary commercial archaeology in the UK*. Unpublished PhD thesis, University of Southampton, Southampton.

Hamilakis, Y. (2004). Archaeology and the politics of pedagogy. *World Archaeology, 36*(2), 287–309.

Hamilakis, Y., & Rainbird, P. (2004). Archaeology in higher education. In D. Henson, P. Stone, & M. Corbishley (Eds.), *Education and the historic environment* (pp. 47–54). London: Routledge.

Hodder, I. (1997). 'Always momentary, fluid and flexible': Towards a reflexive excavation methodology. *Antiquity, 71*, 691–700.

Holtorf, C. (2006). Studying archaeological fieldwork in the field: Views from Monte Polizzo. In M. Edgeworth (Ed.), *Ethnographies of archaeology practice: Cultural encounters, material transformations* (pp. 81–94). New York: Altamira.

Honey, P., & Mumford, A. (1982). *Manual of learning styles*. London: P. Honey.

Jackson, V., & Sinclair, A. (2009). Archaeology graduates of the millennium: A survey of the career histories of graduates. Liverpool: Subject Centre for History, Classics and Archaeology. Retrieved April 24, 2011, from http://www.heacademy.ac.uk/assets/hca/documents/archaeology/Archaeology_Graduates_of_the_Millennium.pdf.

Kolb, D. A. (1984). *Experiential Learning – Experience as the source of learning and development*. Englewood Cliffs, NJ: Prentice-Hall.

Lewis, J. (2006). *Landscape evolution in the Middle Thames Valley: Heathrow Terminal 5 excavations: Volume 1, Perry Oaks*. Oxford and Salisbury: Oxford Archaeology/Wessex Archaeology.

Lucas, G. (2001). *Critical approaches to fieldwork: Contemporary and historical archaeological practice*. London: Routledge

QAA (Quality Assurance Agency). (2007). Subject benchmark statement: Archaeology 2007. Gloucester: Quality Assurance Agency 166 02/07. Retrieved April 21, 2011, from www.qaa.ac.uk/academicinfrastructure/benchmark/honours/archaeology.asp.

Rainbird, P., & Hamilakis, Y. (Eds.). (2001). *Interrogating pedagogies: Archaeology in higher education* (British Archaeological Reports International Series 948). Oxford: Archaeopress.

Stone, P. (2004). Introduction: Education and the historic environment into the twenty-first century. In D. Henson, P. Stone, & M. Corbishley (Eds.), *Education and the historic environment* (pp. 1–12). London: Routledge.

Thorpe, N. (2004). *Student self-evaluation in archaeological fieldwork*. Report for the Higher Education Academy. Retrieved April 21, 2011, from http://www.heacademy.ac.uk/hca/resources/detail/student_self_evaluation_in_archaeological_fieldwork.

Chapter 4
Archaeology for All? Inclusive Policies for Field Schools

Amanda Clarke and Tim Phillips

In·clu·siv·i·ty/inkloō'sivitē/• n. an intention or policy of including people who might otherwise be excluded or marginalized, such as the handicapped, learning-disabled, or racial and sexual minorities (*The Oxford Pocket Dictionary of Current English*).

Excavation is an integral part of learning about and understanding archaeology. A well-trained archaeological workforce is vital to our discipline, and so the task of teaching and learning archaeological field skills is of paramount importance. Are these skills accessible to all, whatever their background and ability? In essence, who do we teach? Is there discrimination of any kind operating in our field schools and what examples of good practice can we take forward? Inclusion in Archaeological Field Schools is discussed, drawing on the results of the Inclusive Accessible Archaeology project run by the Department of Archaeology at the University of Reading in 2005 (Phillips and Gilchrist 2005), and on the experience of running one of the largest archaeological field schools in Britain, at Silchester Roman Town in Hampshire, England. The UK legislation on disability will be briefly reviewed, and how it affects field schools is considered. We will examine the nature of the problem and the challenges that face those of us tasked with training the archaeological workforce.

4.1 The Disability Legislation and Archaeology Students

A number of pieces of anti-discrimination legislation have been passed in Britain over the last 15 years. These relate not just to disability, but also to race, religion, gender, age and sexual orientation. The efforts to eliminate discrimination against people with disabilities are now part of a wider agenda addressing this issue experienced by a wide range of "minority" groups. While some discriminatory issues in archaeology have been discussed in other countries, little has been published elsewhere relating to disability and archaeology. The pieces of legislation that related specifically to disability were the Disability Discrimination Acts (DDA 1995, 2005)

A. Clarke (✉) • T. Phillips
Department of Archaeology, University of Reading, Reading, UK
e-mail: a.s.clarke@reading.ac.uk

H. Mytum (ed.), *Global Perspectives on Archaeological Field Schools:*
Constructions of Knowledge and Experience, DOI 10.1007/978-1-4614-0433-0_4,
© Springer Science+Business Media, LLC 2012

Table 4.1 Types of impairments relevant to field schools

Sensory impairments – sight and hearing
Fluctuating or recurring effects – e.g. ME, epilepsy
Progressive – e.g. motor neurone disease (MND), muscular dystrophy, cancer, HIV, multiple sclerosis (MS)
Organ-specific – asthma, cardiovascular disease, liver and kidney disease
Developmental – autistic spectrum disorders, dyslexia, dyspraxia
Learning difficulties – difficulties with processing the information used for learning
Mental health conditions/diseases, including personality and behavioural disorders
Injuries to the body or the brain

Table 4.2 Types of impairment potentially affecting field school participation

Mobility
Manual dexterity
Physical co-ordination
Continence
Ability to lift, carry or move everyday objects
Speech, hearing or eyesight
Memory or ability to concentrate, learn or understand
Perception of the risk of physical danger

and the Special Educational Needs and Disability Act (SENDA 2001). These have now been superseded by the Equality Act (DDA 2010) which came into force on 1 October 2010 and combines all the anti-discrimination legislation within a single Act of Parliament.

Legally, a person with a disability is defined as someone who has "a physical or mental impairment which has a substantial and long-term adverse effect on their ability to carry out normal day-to-day activities" (DDA 1995). Individuals will have differing degrees of recognised impairments; the main categories are set out in Table 4.1.

It is normal for ability to vary from person to person. To assess the effects of an impairment, several factors need to be taken into account, including the time usually taken to carry out an activity, the way in which an activity is usually carried out, and the environmental conditions such as temperature, humidity, or lighting. The definition of whether an impairment has a long-term effect is "it has lasted at least 12 months and is either likely to last 12 months or likely to last the rest of a person's life. These limits do not apply to people with HIV, MS or cancer, who are defined as "disabled from diagnosis" (DDA 2005). An impairment is considered to have an effect on an individual's ability in one or more of the cases listed in Table 4.2. An individual with an impairment may still be able to carry out normal day-to-day activities, but their mode of activity may be effected by pain or fatigue, or be limited due to medical advice.

Discrimination against a person with a disability is defined as treating them less favourably than other people because of their disability, failure to make any required "reasonable" adjustments for a person with a disability, or any element of victimisation or harassment.

Table 4.3 Archaeology undergraduate students from 16 UK departments with some form of disability or impairment

Disability/impairment	Number	% Disabled students	% All students
Dyslexia	178	63.1	8.6
Hidden disability	43	15.2	2.1
Restricted mobility	24	8.5	1.2
Mental illness	16	5.7	0.8
Hearing impairment	15	5.3	0.7
Asperger's syndrome	3	1.1	0.2
Visual impairment	3	1.1	0.2
Total	282	100.0	13.8

In respect of all aspects of an activity, and the physical features of premises where this activity is carried out, there is a duty in the UK to make reasonable adjustments so as not to place a person with a disability at a substantial disadvantage in comparison with people who are not disabled.

The legislation that most affects the teaching of archaeology in the UK is the Special Educational Needs and Disability Act (SENDA 2001). This makes discrimination against students (and potential students) on the grounds of their disability unlawful. Universities have a duty to make "reasonable adjustments" to ensure that disabled students have full access to all the services they provide. Adjustments must not be "responsive", that is responding to the needs of individuals as they arrive; they must be "anticipatory". These procedures must be in place to provide for the needs of any disabled student.

All Higher Education institutions are required to have explicit policies on disability and provide substantial support and services for students with disabilities. Ensuring inclusion has become an integral part of providing students with an education – it is incorporated at all levels within universities from the policy-making process down to the actual teaching and assessment, and to practicalities such as access to buildings and services. Full participation in archaeological fieldwork training cannot be excluded from this list of activities and services. To date, discussion of inclusion within field schools has concentrated on other under-represented groups such as descendant communities rather than those with disabilities (Baxter 2009; Silliman 2008).

It is difficult to collect accurate figures on the number of students with a disability who are in Higher Education, mainly because all data are reliant on individuals actually declaring their disability; many may choose not to do so. Additionally, some conditions may be undiagnosed. A survey of Archaeology Departments in British universities was carried out by the Inclusive, Accessible, Archaeology (IAA) project in 2005 (Phillips and Gilchrist 2005), which collected data from 16 departments with a total of 2,060 students studying Archaeology at undergraduate level (Table 4.3).

This survey recorded that just under 14% of undergraduate students studying Archaeology had declared some form of disability, although the true figure may be greater for the reasons cited above. The high number of students with dyslexia may reflect the regular screening for this condition now carried out in UK education at all levels. Hidden disabilities include conditions such as Asthma, ME, Diabetes, Heart conditions, Epilepsy and Irritable Bowel Syndrome. Restricted mobility included such conditions as back, knee and joints problems, but only one wheelchair user was identified. Despite the physical "image" of archaeology as a field discipline, a substantial number of disabled students with restricted mobility were choosing this as an undergraduate course of study. Conversely, the low number of disabled students with a visual impairment may represent a perception of archaeology as a very "visual" subject. Despite the difficulties involved in collecting accurate data on the actual number of disabled students studying Archaeology, this survey strongly indicates that a substantial number of them have a disability of one form or another.

A questionnaire survey and interviews were carried out by the IAA project in 2005, and by the Disability and the Archaeological Profession project (DAP) in 2008 (Phillips and Creighton 2010). Some of the information from these is summarised below. Disabled archaeology students were generally quick to praise the practical help, support and the positive attitude and enthusiasm of the staff and support services given by their Higher Education Institutions:

> All the staff have been very supportive and many have shown considerable kindness. I have been on four fieldwork projects since I started at University. I have no 'practical' needs, being able-bodied, but I have recorded my medication on all required forms. Three quarters of fieldwork has been great, one quarter not so good as too much pressure to complete from day one. I am splitting the third year into two parts. The department has supported this decision. (Student: Depression/Anxiety).

Flexibility in how the practical work was organised and carried out on a day-to-day basis was also important for some respondents, and this flexibility included the active support of their peers; one respondent specifically referred to help from other students as well as staff:

> When I did take part in fieldwork, I found that not only the staff but the students were helpful in giving support during excavating. For example, I was allowed to take 5 minute breaks if needed and certain aspects of the excavating that I could not manage, other students freely took over when asked by the staff. (Student: Fused Elbow).

One-to-one tuition at critical times, and the personal communication that this entails, was seen as being of great benefit. The physical act of doing fieldwork as part of a team and being present on an archaeological dig was also seen as effective way of learning in itself for some of the respondents:

> Personally I find being able to place some kind of emotion or visual picture to learning helps assist my memory. I'm a visual learner so practical participating helps. I also find that I need everything to be written down step by step clearly so I can process the information properly. (Student: Dyslexia).

There was also an appreciable number of students who found that they had experienced few or no problems with archaeological fieldwork, and for a couple of students, participation in fieldwork was actually seen as an aid in coping with their condition; "Having to keep accurate notes has helped organise my thoughts" (Student: Dyslexia).

Some individual respondents found difficulty, however, in some of the practical aspects of archaeological fieldwork and the specific problems were often directly related to particular disability/impairments. In many cases, this tended to be the physical demands of the work, "I cannot sustain a repetitive activity for many hours/ days at a time. This causes undue pain and decreases productivity" (Student: RSI, Whiplash, Back Pain, Congenital Hip Disorder). And in another case, "This year concentration and stamina have been difficult; both of these have hindered learning a lot. Pain makes it difficult to participate in fieldwork, but not impossible" (Student: Arthritis, Phobia, Thyroid Problems).

The environmental conditions of excavation such as access to a site and weather conditions were cited as an area of difficulty by a couple of the students. The respondents with dyslexia and similar conditions referred to aspects that are directly related to their personal difficulties:

> In practical work I feel everything has to be spelt correctly, so I feel better if someone helps write things up such as descriptions of finds. I also need to be shown something more than once before I remember how to do it. (Student: Dyslexia, Asthma).

> I forget details easily if I am not doing something; such as, if I have not surveyed for a while I will get confused over back and fore sights and the calculations needed. It will take a lot of revision and people explaining things over and over until I remember and understand again. (Student: Dyslexia).

On only a few occasions were references made to the way sites were being run, for example, "Some Health and Safety issues on site. I felt the standards were unacceptable for me and made me feel uncomfortable" (Student: Dyslexia, Arthritis, Asthma, Upper Limbs Disability, Depression). Also, "There is no provision, or guidelines, on how to accommodate epileptics in academic and developer-funded archaeology, apart from general first aid knowledge" (Student: Epilepsy).

A couple of the students had experienced a number of difficulties. These had become particularly distressing as they had built up from an initial problem which had not been addressed and, as things progressed, the situation had become worse:

> Some members of the Archaeology Department staff have insisted that, even though I have a disability, that I must do more fieldwork in the field (which I have tried to do but have been sent home from the excavation as I was unable to carry out the heavy manual labour required on excavation). This has cost me personally to suffer a loss of confidence and self-worth, as while out in the field other excavators, supervisors and the site director are constantly having to try to find jobs for me to do and this has caused tension as certain staff have thought I was faking my pain and looking for a 'cushy job'. This in turn will affect the personal report the site director does on my contributions to the excavation which is given to my department, and I doubt it will be a good one. (Student: Arthritis, Phobia, Thyroid Problems).

A lack of understanding of the effects and needs of particular disabilities/impairments was cited as a major problem. This included a lack of understanding by other students as well as staff. Some of the students felt that they were being made to look foolish because of ignorance about their condition. This was especially the case where the disability is not particularly "visible":

> People teaching me to draw plans etc. were not very patient when I needed them to explain the process more than once due to my dyslexia. People who train archaeology (sic) need to be made aware of some of the possible difficulties that dyslexic students face and spend more time with them, ideally in a one-to-one situation. (Student: Dyslexia).

LIBRARY, UNIVERSITY OF CHESTER

In summary, the experience of archaeological fieldwork training for several of the respondents to the IAA and DAP surveys had been extremely positive. These came out of a combination of the help they had received, approaching things with a positive attitude, and by making the most of the opportunities they were presented with:

> I took every opportunity to be involved and have had a great experience. On my first excavation I was allowed to stay on for extra time. Have done environmental work, trowelling and other jobs including dendrochronology and the tree recovery team (submerged forests). Have a wide range of experience through the University as a student. I need to work at my own pace, difficulties if I had to work at commercial rate, but I was able to specialise. I have been encouraged by complimentary comments from academics and professionals and fantastic support from people on excavations. It has helped me rebuild my life [after the accident], a very positive experience. (Student: Dyslexia, Arthritis, Asthma, Upper Limbs Disability, Depression).

Another respondent related how they purposefully reviewed each situation they found themselves in, and then discussed it with their tutor. They considered that effective communication by themselves about their condition and needs led to a greater understanding by the staff of what was required:

> I have always made a point, whether on a dig or during study, of letting the Tutor/Leader know if I am finding something difficult on a particular day, or if I predict that some activity requiring particular skills (i.e. penmanship and drawing in my case) may be challenging. I continue to believe this is the best way to manage my programmes and MS. (Student: MS).

Finally, two of the students offered advice based on their experiences. One was directed to other disabled students and emphasised making the most of opportunities, "Have faith in yourself that you can do the course. Listen to what people tell you and watch what people are doing. Have the courage to ask questions" (Student: Dyslexia).

The second piece of advice concerned project guidelines:

> In the main, disabled people dislike being 'nannied'. Please do not over-regulate, this always achieves the opposite of what is intended, particularly where the regulation is introduced with the best of motives. One only has to look at the school trips/risk analysis industry to observe the pitfalls of such an approach. (Student: MS).

4.2   The Silchester Context and Challenges

The issues that face all field schools – in Britain and elsewhere – have been confronted at some point during the 15-year (and continuing) lifetime of the Silchester Field School, an undergraduate research and training excavation run by the Department of Archaeology at the University of Reading. Over the 84 weeks of this project, we have developed a policy of inclusion for the field school (Stewart et al. 2004).

Fig. 4.1 The Silchester Field School: looking south-east over the excavation trench in Insula IX (copyright Department of Archaeology, University of Reading)

Silchester, the site of the Roman town of *Calleva Atrebatum*, lies halfway between Reading, Berkshire, and Basingstoke, Hampshire, in the UK. The town has been the subject of systematic archaeological investigations intermittently since Victorian times, and its status as an undeveloped rural site, preserved as such after its abandonment in the fifth/sixth century AD, has guaranteed continued archaeological interest (Boon 1974; Clarke and Fulford 2002). Through investigation of a large part of one of the town's *insulae* (city blocks), the Silchester "Town Life" project aims to explore the origins, development, decline and eventual abandonment of part of *insula IX* and to learn as much as possible about the changing life of the town through its history.

A large excavation area of 3,025 m^2 was opened in 1997 and was able to accommodate the entire first-year intake of archaeology undergraduates from the University of Reading, and thus provide, for the first time, equality of opportunity in fieldwork training for the Department. It is now anticipated that the fieldwork phase of the project will continue at least until 2015, and in 2010, over 80 students from the University of Reading and over 127 participants from elsewhere attended for between 1 and 6 weeks of field excavation (Fig. 4.1).

Over the years, the training has evolved into the Silchester Field School which now teaches archaeological field techniques not just to Reading undergraduates, but also to students from all over the world, including A-level and mature students, and

Fig. 4.2 Hands-on teaching and learning in the excavation trench at Silchester Roman Town, Insula IX (copyright Department of Archaeology, University of Reading)

to interested amateurs (Clarke 2010). The nature of on-site training has of necessity changed and developed alongside an increase in participant numbers and the delivery of training to such large and diverse groups of people. Correspondingly, assessment has become an important and transparent part of the process.

The "Silchester Experience" is now offered to and enjoyed by nearly 300 people a season, and it operates from the heart of the Roman town. Students from the University of Reading spend either 4 or 2 weeks at Silchester, depending on whether they are Single or Joint Honours students. The training provided is a mix of formal introductory sessions with actual hands-on experience, including interactive site tours, introductory sessions on Roman finds and the handling of artefacts, and skill-gaining sessions on excavation techniques, and the taking of environmental samples. These sessions are later supplemented by more detailed sessions on science in archaeology, stratigraphy, archaeological planning and survey, finds drawing, archaeological reconstruction, and archaeological photography. All newcomers are assigned to a professional supervisor and everyone has the opportunity to try their hand (under close supervision) at all aspects of excavation and field recording. Each participant is given the opportunity (again under constant supervision) for one small part of the site and taught the entire process of excavation, recovery and recording (Fig. 4.2). Integral to the Silchester experience is the chance to work as part of a

Table 4.4 Silchester Field School 2009, 2010 and 2011: number of university of reading first-year students with disclosed disabilities

Disclosed disability	Number		
	2009	2010	2011
Dyslexia/dyspraxia	12	9	10
Asthma	1		
Epilepsy	1		1
Asperger's syndrome	1	1	
Bipolar disorder + dyspraxia + scoliosis	1		
Impaired vision in one eye		1	
Attention deficit disorder		1	
Obsessive compulsive disorder		1	
ME			1
Chronic fatigue syndrome			1
Impaired hearing			1
Congenital heart block			1
% Total of all part 1 students	32%	22%	Add here

small supervised team within larger groups of all ages, backgrounds and abilities, towards a common research aim.

A major challenge for those faced with setting up a seasonal compulsory excavation is how to work with participants of differing abilities, particularly those who have declared a disability. For example, during the 2009 season of excavation at Silchester, 32% of the attending University of Reading first year students declared a disability; in the 2010 season this dropped to 22%, and the number who have declared a disability for the academic year 2010/2011 has further dropped to 18% (although it is likely to rise for the start of the next Silchester season, based on previous trends). Therefore, an average of 25% of all the assessed students at Silchester has declared a disability (Table 4.4).

The Silchester Field School is also open to outside participants, and in the 2009 season, 28% of outside participants declared a disability, and in the 2010 season 23% declared a disability (Table 4.5).

The types of disability vary – among the students dyslexia/dyspraxia is by far the most common (on average 70% of the disabilities declared). On one hand, this could be the result of intensive testing for dyslexia in UK higher education – or even the possibility that archaeology as a discipline attracts higher numbers of dyslexic students. As one respondent to the IAA project questionnaire put it:

> When I was first choosing which subject to study, I was told that archaeology was a dyslexic-friendly subject and I guess that is the reputation it has. This is probably true! (Student: Dyslexia).

Mental disabilities are often not openly declared. The number of participants on antidepressants, for example, may only be disclosed unofficially during the course of the field season in response to a particular situation. Finding ways of including and motivating students with mental health issues who are living away from home and camping in a field is a challenge indeed. How do we as academics cope with the

Table 4.5 Silchester Field School 2009 and 2010: number of other participants with declared disabilities

Disability	Number in 2009	Number in 2010
Asthma	7	7
Eczema/allergies	1	2
Arthritis	2	2
High blood pressure	3	2
Tendonitis	0	1
Hip replacement	1	1
Knee replacement	1	0
Shoulder injury	1	1
Back problems	4	4
Anterior cruciate ligament repair	1	0
Narrowing of pulmonary arteries	0	1
Severe hay fever	1	1
Diabetes	2	2
Asperger's syndrome	2	1
Thyroid problems	3	1
Depression	2	1
Schizophrenia	1	0
Epilepsy	1	0
Chronic fatigue	1	0
Registered disabled – limited movement	1	0
Hearing problems	1	0
Anaemia	0	1
Cardiac irregularities		1
HIV	1	0
% Total of all participants	28%	23%

600% increase in a decade of the number of UK students declaring a mental-health problem? In 1994–1995, just five students in every 10,000 declared an issue with their mental health. By 2004–2005, the figure had risen to 30 in every 10,000, according to figures from the Higher Education Statistics Agency (Times Higher Education 10 April 2008).

The disabilities disclosed by outside participants are much more varied than those of university students, and include muscle problems, back, shoulder, knee injuries, breathing problems, and allergies – and, again, problems with depression. Not surprisingly, given the wide age range from 16 to over 70, and with between 30 and 36% of outside participants being over the age of 30, there is a bias towards illnesses common to older age groups. Generally, the disabilities disclosed by outside participants are more manageable as the more mature participants are often able to control their own conditions – and many of them involve more practical issues, such as the inability to lift buckets of heavy spoil.

4.3   One Approach to Making Archaeological Fieldwork Inclusive: The Inclusive, Accessible, Archaeology Project (IAA)

In response to the current legislation, the onus is on British universities to ensure that all students have access to the practical aspects of the degree courses they offer and to make anticipatory, reasonable adjustments to ensure that no individual is excluded for reason of their disability. This can be a major challenge because it is practically impossible to anticipate the specific needs of every individual student who may, or may not, be present on fieldwork training. Being placed in a totally new environment, the individual student may not even be able to anticipate what their needs may be. The reality is that not everyone can do everything, and this will certainly be the case with some of the practical tasks undertaken in archaeology. All individuals possess different levels of ability whether they are disabled or non-disabled. For some people, their level of ability may mean that they cannot partici-pate fully in a particular task. However, deciding whether this is the case should be based on their individual abilities, not on them being labelled as a "disabled" person. Inclusion is therefore not just about disability; it is about providing access for all students.

Two aspects of the ability to participate in archaeological tasks can be identified: physical and cognitive. In order to identify these physical and cognitive demands, the IAA project listed the major tasks that UK Archaeology departments teach and assess as practical work. These include: excavation and planning, surveying, geo-physical and field survey, environmental sampling, and the processing of artefacts. The physical and cognitive demands of the various aspects of the individual tasks were then characterised. This involved close cooperation with occupational thera-pists and professional access consultants to observe the various archaeological tasks under controlled conditions. As well as the main activities involved in excavation, such as using a wheelbarrow (Fig. 4.3) and excavating with a trowel, other tasks were also included, such as methods of recording; discerning stratigraphy through vision, colour, texture and touch; and climbing in and out of trenches. A number of everyday activities that replicate the archaeological tasks were identified and visual acuity tests developed. From this, a prototype tool kit was devised. In tests under controlled conditions with 20 disabled and non-disabled volunteers, the tool kit was refined. The final version was then distributed in field trials on three training excava-tions with around 120 disabled and non-disabled students. The tool kit was named the Archaeological Skills Self-Evaluation Tool kit (ASSET) and is available as an online resource at: http://www.britarch.net/accessible. This can be used or adapted by other to fit their particular circumstances, as it is a generic framework for evalu-ating skills and abilities.

The tool kit has been designed for use primarily by people who have little or no experience of archaeological fieldwork. Prior to fieldwork, an individual can use this resource to identify their potential ability to carry out various tasks successfully by answering a series of questions about everyday activities that replicate the

Fig. 4.3 Using a wheelbarrow at Silchester Roman Town, Insula IX excavations (copyright Department of Archaeology, University of Reading)

archaeological ones. These can be answered at different levels of difficulty, thus identifying where support or adjustments may be needed. After fieldwork has been completed, the student can evaluate their practical fieldwork abilities, again at various levels. They can also assess how well they have performed in carrying out the various activities. As abilities and skills are evaluated at different levels of attainment, the tool kit can be used on subsequent occasions to track student development.

The research carried out by the IAA project indicates that the inclusion of disabled students in fieldwork training has been successful in cases where their abilities and limitations were known and fully understood beforehand. Problems arose where this was not the case. An important part of the process is a review of provisions and procedures after participation, and the post-fieldwork evaluation provided by the tool kit can help with this procedure and, because it incorporates the dynamic nature of ability, any future adjustments can be changed or adapted to suit the individual student. The tool kit can also be used before embarking on a degree course in archaeology as it can give prospective students an idea of their potential abilities.

The Silchester Field School is set up to embrace, encourage and inspire beginners. The infrastructure is such that we are able to uphold the ideal that beginners of all ages, all backgrounds, all abilities, and all motivations can slot into a large ongoing project in which they can thrive and develop a variety of skills. The Silchester credo is that archaeological excavations can provide an experience for everyone, no matter what their ability. All participants have differing abilities; some are better with their hands, some with their eyes; the ability to "see" three-dimensionally is not something only the able-bodied can do.

The learning objectives are based on attainment of a large number of skills which are presented in the Field School Handbook (Clarke 2010), and students are expected to make constant reference to this. This allows them the opportunity not only to monitor the different archaeological skills gained while on site, but also to assess critically their own site performance. Part of the assessment is recognition of and reflection on all skills gained on-site, and a self-assessment essay is an integral part of the experience for first-year Reading University students. The assessment is broad-based and takes into account the individual's performance on a day to day basis, whatever tasks they are undertaking. The assessment sheets are designed to reflect the variety of cognitive and physical skills gained during excavation. After the field school, everyone has the opportunity to carry the skills gained – and recognised – into their second year as part of their personal development planning for developing employability.

The Silchester Town Life *Insula IX* project has developed a large-scale pedagogic and research project which facilitates, encourages, and in turn benefits from broadening participation. At the start of each season, every participant is asked to declare his or her own particular areas of challenge, and this information is communicated to each and every supervisor on site. We recognise that everyone has differing abilities and we make provision for these. Each participant is encouraged to take part in all aspects of work on site, from wielding a trowel or a pickaxe, to describing a context, drawing a plan, entering data onto the project's database, working with artefacts or scientific samples, through to meeting and greeting visitors to the excavation and giving site tours. We recognise that we all have different abilities – and different challenges – and if we start from the premise that excavations are for all, there is something for everyone; no one is excluded. Students with physical challenges – from back injuries, to arthritis, to lower limb paralysis – are encouraged to test their own limits – and to set their own requirements.

Mentors are made available to carry spoil and empty wheelbarrows; ramps are constructed to allow easy access to trenches – and we start from the belief that anything is possible. Much of assessment on site is about attitude, communication, passion and interest – and demonstrating this is the key to benefiting from the fieldwork experience. Other physical challenges, such as partial sight, can also be encompassed. One year, a partially sighted student was able to excavate small areas using touch (Fig. 4.4), and with the help of a mentor. Building this into the experience means that those acting as mentors also develop and strengthen their own abilities at the same time.

Fig. 4.4 Archaeology by touch: a visually impaired student excavating at the Silchester Field School, Insula IX, Silchester Roman Town (copyright Department of Archaeology, University of Reading)

An excavation tests and showcases so many abilities – from oral communication to working with numbers and figures. For example, initially apprehensive students have wholeheartedly welcomed interaction with visitors as part of their assessment – they have recognised the enjoyment and value of communication and explanation and have in the process grown in confidence. Those who struggle with the written word may blossom when it comes to giving site tours to interested visitors. Indeed, this is one of the most successful aspects of the Field School – most visitors comment on the high standard of interaction with the students and the explanations they receive.

A field school can provide a comfortable environment in which to explore the variety of skills each and every one of us has. The recognition that archaeological fieldwork is not just about digging is the starting point to providing a truly inclusive experience. Regardless of disability or ability, everyone can benefit and learn (Fig. 4.5).

At the outset, it is important to have all paperwork in place; these include signed disclaimers and medical forms. The project risk assessment must be comprehensive, particularly in relation to the varied abilities of the participants and the measures that may need to be put in place to aid those with physical challenges; it is important

Fig. 4.5 Archaeology for all:
recording deposits at the
Silchester Field School,
Insula IX, Silchester Roman
Town (copyright Department
of Archaeology, University
of Reading)

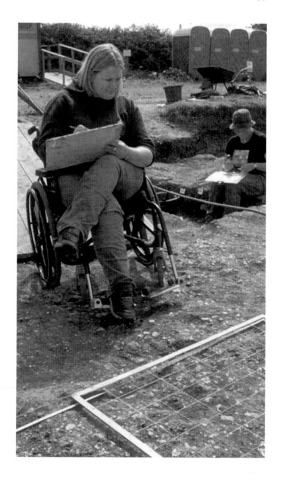

that the archaeology (and the campsite) is made as accessible as possible. The risk
assessment should also discuss those with mental issues; it is important to gain the
backing and the support of the University department and the students, personal
tutors or advisors, and to communicate in advance about any potential problems.
For all participants, the learning environment is structured so that all field school
staff are aware of the inclusive nature of the project and of the different abilities of
all those in their care. Above all, everyone should understand the range of site tasks
available; if those with back problems cannot carry a bucket of spoil then there are
choices: a physical mentor can step in, or the participant can be given the opportunity to spend part of the day doing a less physically demanding task. Flexibility is
the key to inclusivity, and assessment should include the recognition of flexible
ability.

4.4   Gender and Age

Discrimination is not just about disability, but can also be about age, gender, sexuality, or ethnic origin and can extend into the wider world of work. For example, it is not unknown for employers (and indeed some excavation directors) to be predisposed to recruiting physically fit male participants for certain types of archaeological excavation. Equally, the image of a female-dominated finds hut persists – are our training excavations helping to prolong the lifetime of these stereotypes?

Of interest is whether such diversity can present a barrier to learning in any way – and do our teaching methods need to take this into account? Any university field team will be made up of a range of individuals; do we need to adapt our teaching and learning methods to reflect this? Perhaps, as with disability, the premise remains the same: we all have different abilities regardless of (and perhaps because of) sex, age, race and gender. An archaeological "toolkit" based on self-reflection and the challenge of identifying our own unique abilities is one step towards developing a cognisant archaeological workforce.

In the 2008 season at Silchester, 58% of the first-year students were female. In 2009, 68% were female, and in 2010, 61% were female. The gender distribution of non-student participants at Silchester is similar; in 2009, 58% were women; in 2010 59% were women (see also Chap. 6). Of interest is to see how these figures carry on into employment, and Aitchison's profile of archaeological jobs (Aitchison and Edwards 2008) shows that only 41% of archaeologists at the time of the survey were female. Figures for all employees in the UK for the period of the survey were 46% female. The figures from Silchester highlight a larger number of female archaeology undergraduates, and a greater number of women over men being trained in field archaeology; this statistic is not being borne out in the archaeology employment figures (Table 4.6). Women are clearly under-represented in the archaeological profession in the UK and the conclusion could be that some discriminatory factors are operating at employment level, or that women select non-fieldwork careers within archaeology and the heritage sector.

At Silchester, equality of field training, whatever the gender, is the norm. All participants are required to spend time both in the finds hut and in lifting buckets of spoil on site. However, it is certainly true that the employed finds supervisors at Silchester have all been women since the project began in 1997 (see also Chap. 7). Aitchison's profile of archaeological employment highlights the fact that finds posts still tend to be the domain of women (Aitchison and Edwards 2008) and women overall tend to be better represented in artefact research than in other areas of archaeology; could it be that more women are "better" at finds or "naturally" more interested, a debate aired by Swift (2007)?

Staff employed at Silchester in 2009 and 2010 were 61% female – which very much reflects the gender proportion studying Archaeology and being trained in archaeological fieldwork – but bucks the employment trend nationally. Aitchison's survey reports that only two fifths of those employed in field investigation and

Table 4.6 Gender balance in archaeology and the UK working population (data from Aitchison and Edwards 2008)

	Archaeologists		UK working population (millions)	
Female	1,013	41%	13.42	46%
Male	1,432	59%	15.80	54%
Total	2,445	100%	29.22	100%

Table 4.7 Gender by individual's principal role – UK archaeologists (data from Aitchison and Edwards 2008)

	Male		Female		Total	
Field investigation and research services	1,025	61%	659	39%	1,684	100%
Historic environment advice and information services	216	55%	176	45%	392	100%
Museum and visitor/user services	44	37%	76	63%	120	100%
Educational and academic research services	111	60%	74	40%	185	100%
Archaeological management	33	58%	24	42%	57	100%
Total	1,429	59%	1,009	41%	2,438	100%

research are female – but in museum and visitor/user services, almost two thirds of archaeologists are female (Table 4.7).

In the interviews conducted by the IAA project in Reading, one mature student summed up the challenges fieldwork training and assessment posed for the older trainee most succinctly:

Assessment: physical frailty and *anni domini*. (Angie: MS).

At Silchester, not surprisingly, between 85 and 95% of the first-year students are aged 19–21, and less than 10% are over age 30. In 2009, only 2% of students were over 60. Of Silchester's outside participants, the age range is greater; in 2009 47% were aged 15–19 years of age; 23% were aged between 20 and 29 and 30% were over 30 (4% over 70). In 2010, 40% were aged 15–19, 22% were aged 20–29 and 36% were over 30 (including 2% over 70). Thus, a wide range of ages participate, from 14-year-olds carrying out work experience to 78-year-old mature students. The training opportunities offered by the project are the same for all; we rely on the broad range of skills on offer and the flexible, inclusive nature of the project to provide archaeology for all. We encourage any participant with age-imposed requirements, from the very young to the more mature, to assess their own abilities prior to beginning work on site. This suggestion of flexible working options can then be reviewed over the course of the field training. The aim is to embrace diversity and use peoples' rainbow abilities to the advantage of the project specifically, and to the discipline in general.

4.5 Conclusions

Departments of Archaeology in the UK are required by law to make the educational experience of archaeology as inclusive as possible. Archaeological excavations are integral to a study of archaeology; they provide an inclusive environment in which to gain and practice a variety of skills. Overall the response of archaeology students to the excavation environment is positive; if both teachers and learners have the appropriate attitude, a high degree of inclusivity can be achieved. The Silchester Field School is an example of a large, long-running research and training excavation which has been able to focus on achievable skills rather than lack of ability. We are not without our failures however, but over 14 seasons of excavation and $c.1400$ students and $c.2800$ outside participants, less than 2% have dropped out for reasons of ill-health. These range from those who struggled to come to terms with the communal nature of an archaeological excavation, to the physical challenges of spending long hours outdoors. Unfortunately, the reality and discomfort of a tent by night and an archaeological trench by day is sometimes not made up for by the thrill of discovery. These cases, however, are few and far between.

An interviewee for the Inclusive Accessible Archaeology project with Multiple Sclerosis provides a view of the experience of being disabled, and perhaps points the way forward:

> I think with sensitivity, and being aware that we all have disadvantages of one sort or another, that archaeology could be a lot more inclusive than it is. Before I started, I had the view of a young, fit and healthy image. Not so much an image problem, more of an image factor. I am sure that if the idea that we cannot all do everything could be got across, it would be a lot better. That is being a human being, not a disabled person. If that idea could be developed, I do not see why archaeology cannot be inclusive.

References

Aitchison, K., & Edwards, R. (2008). *Archaeology labour market intelligence: Profiling the profession 2007–08*. Reading: Institute for Field Archaeologists.

Baxter, J. E. (2009). *Archaeological field schools. A guide for teaching in the field*. Walnut Creek: Left Coast Press.

Boon, G. C. (1974). *Silchester: The Roman town of Calleva*. Newton Abbott: David & Charles.

Clarke, A. (2010). *Silchester field school handbook for students 2010*. University of Reading: Department of Archaeology.

Clarke, A., & Fulford, M. (2002). The excavations of *insula IX*, Silchester: the first five years of the 'Town Life' Project, 1997–2001. *Britannia, 33*, 129–66.

DDA (Disability Discrimination Act). (1995). London: HMSO. http://www.legislation.gov.uk/ukpga/1995/50/contents. Last consulted 21 April 2011.

DDA (Equality Act). (2005). London: HMSO. http://www.legislation.gov.uk/ukpga/2010/15/contents. Last consulted 21 April 2011.

Phillips, T., & Creighton, J. (2010). *DAP: Disability and the Archaeological Profession. Employing People with Disabilities: Good Practice Guidance for Archaeologists*. IfA Professional Practice Paper no. 9. Reading: Institute for Field Archaeologists.

Phillips, T., & Gilchrist, R. (2005). *Disability and Archaeological Fieldwork Inclusive, Accessible, Archaeology Phase 1 Report*. Available at: http://www.hca.heacademy/access-archaeology/inclusive_accessible. Last consulted 21 April 2011.

SENDA (Special Educational Needs and Disability Act). (2001). London: HMSO. http://www.legislation.gov.uk/ukpga/2001/10/contents. Last consulted 21 April 2011.

Silliman, S. W. (Ed.). (2008). *Collaborating at the Trowel's edge. Teaching and learning in indigenous archaeology*. Tucson: University of Arizona Press.

Stewart, R., Clarke, A., & Fulford, M. (2004). Promoting inclusion: facilitating access to the Silchester 'town life' project. *World Arch, 36*(2), 220–235.

Swift, E. V. (2007). Small objects, small questions? Perceptions of finds research in the academic community. In R. Hingley & S. Willis (Eds.), *Roman finds: Context and theory* (pp. 18–28). Oxford: Oxbow.

Chapter 5
Archaeological Field Schools and Fieldwork Practice in an Australian Context

Sarah Colley

5.1 Introduction

Most archaeological field schools are run by universities for students in collaboration with external organisations. Approximately 21 of Australia's 37 universities currently teach some Archaeology (Australian Archaeological Association 2010a), and about 11 offer a full major, fourth year Honors and postgraduate degrees in Archaeology (Table 5.1). Some universities run several assessed archaeological field schools and significant field-based teaching, while elsewhere these are offered irregularly if at all. A brief history of fieldwork teaching at University of Sydney, and examples from other universities, allows discussion as to why the provision of field schools varies significantly between institutions and locations across Australia.

Some Archaeology has been taught in Australian universities since 1948. Before the 1960s, this was mainly "Near Eastern" and Classical Mediterranean Archaeology and Egyptology. These remain popular with Australians and continue to influence university curriculae and field opportunities (Balme and Wilson 2004). From the 1960s and 1970s, university courses expanded to include prehistory, historical and maritime archaeology in Australasia, the Pacific and elsewhere. Universities flourished into the 1980s and new departments were established (e.g. Fredericksen and Walters 2002; Hall 1980). Curriculae changed over time (Feary 1994; Frankel 1980), and most Australian universities now offer a mix of teaching about archaeology and heritage in different regions, including Australasia, with variable focus on prehistory, text-aided/historical and maritime archaeology as well as cultural heritage studies, theory, methods and practice (Australian Archaeological Association 2010a).

S. Colley (✉)
Department of Archaeology, School of Philosophical and Historical Inquiry,
University of Sydney, Sydney, NSW, Australia
e-mail: sarah.colley@sydney.edu.au

H. Mytum (ed.), *Global Perspectives on Archaeological Field Schools:*
Constructions of Knowledge and Experience, DOI 10.1007/978-1-4614-0433-0_5,
© Springer Science+Business Media, LLC 2012

Table 5.1 Australian universities offering major degree programmes in Archaeology 2010–2011

Name and location of University	Organisational Unit 2010
Australian National University (ANU), Canberra, Australain Capital Territory	School of Archaeology and Anthropology
Flinders University, Adelaide, South Australia	Department of Archaeology, School of Humanities
James Cook University (JCU), Townsville, Queensland	Department of Anthropology, Archaeology and Sociology, School of Arts, Education and Social Sciences
La Trobe University, Melbourne, Victoria	Archaeology Program, School of Historical and European Studies, Faculty of Humanities and Social Sciences
Macquarie University, Sydney, New South Wales	Department of Ancient History, Faculty of Arts
Monash University, Melbourne, Victoria	Centre for Archaeology and Ancient History, Philosophical, Historical and International Studies, Faculty of Arts
	School of Geography and Environmental Science, Faculty of Arts
University of Melbourne, Melbourne, Victoria	Centre for Classics and Archaeology, School of Historical Studies
University of New England (UNE), Armidale, New South Wales	Archaeology and Paleoanthropology, School of Humanities
University of Queensland (UQ), Brisbane, Queensland	Archaeology Programme, School of Social Science, Faculty of Arts
University of Sydney, Sydney, New South Wales	Department of Archaeology, School of Philosophical and Historical Inquiry, Faculty of Arts
University of Western Australia (UWA), Perth, Western Australia	Archaeology, School of Social and Cultural Studies, Faculty of Arts, Humanities and Social Sciences

Other Australian universities not listed also teach some archaeology. Information from university websites November 2010 and Australian Archaeological Association (2010a)

The balance between local and overseas research focus varies significantly between institutions. Most key Australian universities conduct major archaeological fieldwork outside Australia, in the Pacific, South East and Western Asia, China, Egypt, the Mediterranean and elsewhere (Table 5.2), in addition to Australian-based projects. Some universities focus strongly on Australian and regional archaeology (e.g. Flinders University, University of Western Australia), while others do little if any (e.g. Macquarie University, University of Melbourne). As discussed below, the chosen regional research focus influences teaching of field schools by universities.

Universities underwent major reconfiguration in the 1990s following government funding cuts and growth of mass higher education. This process continues to affect professional education and training (e.g. Colley 2004; Colley et al. 2005). Completing fourth year Honors in Archaeology was traditionally the basic minimum requirement to work in the profession. This is now questionable due to significant changes to university degree programmes and curriculae (Beck and Balme 2005). Growing

Table 5.2 Assessed field schools offered to undergraduates 2010–2011 listed on university websites

University	Major regional research focus	Archaeological field schools 2010–11
ANU	Pacific, South East & North Asia, Australia	1 × 2–3 week field school (Philippines) out of semester + post-excavation work on campus. Costs $1200–1600
Flinders	Australia, Pacific (+USA, South America)	ca. 10 × 1–2 week or shorter field schools in Australia (+ Pacific & S. America) mainly out of semester. Costs ca. $500–$3,000+depending on location, duration and if student is local, interstate or overseas. Degree credit also offered for approved archaeological fieldwork experience
JCU	Australia; South East Asia	1 × 1 week maritime archaeology or rock art field school offered alternate years in Australia (Queensland). Out of semester. Unspecified costs apply
La Trobe	Australia, South East & North Asia, Mediterranean, Africa	No formal field schools. Degree credit offered for minimum 4 consecutive weeks of approved archaeological fieldwork experience. Offered during and out of semester
Macquarie	Egypt, Mediterranean	1 × 6 weeks participation in a University affiliated field project/excavation in Egypt or another approved archaeological excavation in the Mediterranean or W. Asia. Out of semester. Costs ca. $4000
Monash	Egypt, Australia, Pacific	1 × 3 week field visit/field school in Italy out of semester. Cost ca. 5900 1 × 2 week field school in Australia (Victoria). Schedule not listed. Not offered 2011. Unspecified costs apply
Melbourne	Mediterranean, West Asia, Egypt	No assessable field schools
UNE	Australia, South East Asia, Pacific	ca. 3 × 6 day field schools in Australia (NSW). Some not offered 2011. Geophysical survey and other field schools open to archaeology students. Unspecified costs apply
UQ	Australia, Pacific, Africa	No assessable field schools. Student handbook notes that several units of study include fieldwork
UWA	Australia, South East Asia	At least one assessable field school for undergraduate students. Not offered 2011. Degree credit offered for approved archaeological fieldwork experience (Arts Practicum). No information about costs available
Sydney	South East, West & Central Asia, Mediterranean Australia, Pacific	No assessable field schools

Some universities offer additional field schools that are not assessed or are run for postgraduate coursework students

professional employment opportunities in Australian archaeology and cultural heritage management beyond the university system followed the introduction of heritage legislation from the 1970s (Colley 2002). A survey by Ulm et al. (2005)

demonstrated that over 70% of archaeologists worked in the heritage management industry compared to a minority in universities or museums.

Field schools and opportunities for volunteer participation in field archaeology in Australia are also organised by government heritage agencies in conjunction with archaeological societies or heritage consultancy companies. A notable pioneering example was the VAS (Victoria Archaeological Survey) Summer Schools in Archaeology that ran between 1974 and 1980 and were open to volunteers and students (Coutts and Wesson 1980). Such a scheme would now be much more difficult to implement due to changed heritage laws and policies, insurance and occupational health and safety (OH&S) requirements.

Some organisations set up by and for members of the public interested in Australian archaeology have run fieldwork projects and excavations (e.g. Archaeological and Anthropological Society of Victoria; Canberra Archaeological Society), while rock art recording has long been a strong area of avocational involvement (e.g. Colley 2002: 140–148). However, compared to the USA and UK, there is far less avocational or "non-professional" interest in local archaeology in Australia. More people want to travel overseas to visit famous archaeological sites and museums and experience ancient artworks and places for themselves (Colley 2007a: 34–35).

A comprehensive suite of laws and policies was introduced from the 1960s onwards to protect and manage most Australian heritage places. Issue of archaeological excavation and fieldwork permits is now regulated, and it is not usually possible to direct an archaeological excavation without appropriate qualifications and experience (Smith and Burke 2007: 124–163). This usually prevents amateurs and students running excavations without professional supervision. The dictates of legislation and policy vary significantly between jurisdictions (e.g. state, territory or federal government) and types of place and practices (e.g. Indigenous Aboriginal, historical and maritime archaeology).

Depending on the location and type of project, archaeologists may need to seek permission from multiple government agencies and negotiate with several community groups and stakeholders to conduct fieldwork. There is no national standardisation on issue of fieldwork permits; this depends entirely on circumstances and the individuals involved and is usually a state government responsibility. Heritage managers and other stakeholders are not always sympathetic to large-scale student participation in excavations as training exercises and this influences the issue of permits for field schools.

State heritage agencies sometimes encourage and support volunteer participation on suitable developer-funded excavations of urban historical archaeological sites (e.g. Sydney's Big Dig and Melbourne's Little Lon). Such opportunities are sporadic and the commercial imperatives, as well as OH&S and insurance costs, impede participation of inexperienced people in most cases. The Canberra Archaeological Society, sponsored by the ACT government, has excavated historic sites and offers training to volunteers. Adult education classes sometimes offer archaeological fieldwork training on historic sites. The Australian Institute for Maritime Archaeology (2010) regularly runs internationally recognised field training in maritime archaeology, in conjunction with the UK Nautical Archaeology Society, which beginners may attend. Field schools and training excavations outside Australia are also popular with students and volunteers.

There are fewer opportunities to participate in fieldwork on Indigenous Aboriginal sites compared to colonial period "European" sites. Research access to Indigenous places always involves negotiating community consent and must comply with cultural protocols. Participation of inexperienced volunteers from outside the community may be inappropriate (see Smith and Burke 2007: 178–193). For example, some Indigenous places are gendered and only open to men or women. Hall et al. (2005: 48) noted that some Indigenous communities in Queensland did not want untrained students working on their important places and this restricted university field school options. Fieldwork on Indigenous sites and in locations away from major urban centres tends to involve fewer personnel and presents less scope for volunteer participation. Fieldwork hazards and OH&S issues are a concern everywhere, and especially in more remote locations and on maritime sites where stringent safety standards are enforced (Smith and Burke 2007: 88–106).

5.2 A Brief History of Fieldwork Teaching at University of Sydney

University of Sydney is not typical in its approach to fieldwork teaching (Table 5.2). However, Sydney has been subject to broad trends experienced everywhere, and comparing Sydney to other institutions is instructional. Much has been published about the history of Archaeology at University of Sydney (Eslick and Frankel 2006; Ireland and Casey 2006; Megaw 2000) and the author has personal experience of changes to teaching over time.

The current department of Archaeology, located in the School of Philosophical and Historical Inquiry in the Faculty of Arts and Social Sciences, has nine full-time teaching and research academic staff, plus research fellows, honorary associates, support staff and others employed by several affiliated organisations on and off campus. The department employs more full-time academic staff than others in Australia but is not the largest overall (Fig. 5.1). Most Australian universities rely on teaching contributions from research staff, affiliates, and casual lecturers and tutors. Accurate information about actual staffing levels is rarely publicly available and these change from year to year.

Sydney offers some fee-income generating postgraduate and Winter and Summer School courses, but the main teaching focus is on the undergraduate major in Archaeology as part of Bachelor of Arts and other combined degrees, fourth year Honors and research Masters and PhDs. Currently, 165–200 students enrol in each of two first year units of study, with about 50% taking both. Senior units typically attract 20–40 enrolments with some lower (6–15) and others significantly higher (60–75). As undergraduate units with fewer than 20 students now make a loss, university policy is to cancel these in future unless subsidy can be justified. Archaeology

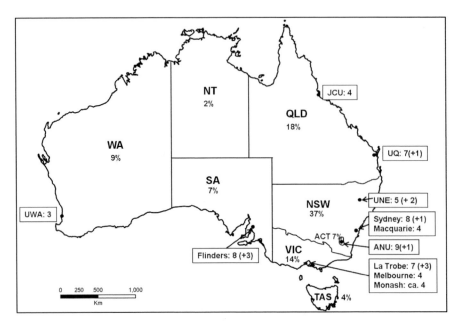

Fig. 5.1 Location of universities listed in Table 5.1 with approximate number of full-time academic teaching staff plus technical and administrative support staff employed in each archaeology department(s). Percentage of professional archaeologists employed by state follows Ulm et al. (2005) where $n = 292$. Figure produced by Annika Korsgaard

attracts fewer undergraduates than other major humanities disciplines in the same School, but the Department has a comparatively high number of Honors and post-graduate research students and a strong staff research profile. The current geographical focus of departmental research is mainly outside Australia, e.g. in Iran, Greece, Cyprus (Fig. 5.2), Southern Italy, Cambodia, China, Central Asia, Jordan, Solomon Islands, Papua New Guinea with just a few Australian projects, mainly in the local Sydney region.

Separate sub-disciplinary organisational units ("Prehistory" and "Near Eastern", "Classical" and "Historical Archaeology") were previously independent, and a combined departmental structure and blended curriculum is a recent development. Sydney inherited a legacy of disagreement about the aims and focus of different kinds of archaeological practice from strong personalities who pioneered the sub-disciplines of Archaeology at the University in the heady days of the 1960s and early 1970s (O'Hea 2000; Megaw 2000; Eslick and Frankel 2006). Significant private funding and community support for Classical Mediterranean and Near Eastern Archaeology have played a key role in perpetuating different departmental "cultures".

Institutions and sub-disciplinary cultures everywhere perpetuate their own traditions of fieldwork practice, student training, mentoring and academic patronage (e.g. Ucko et al. 2007). Provision of field schools and formal teaching of field methods

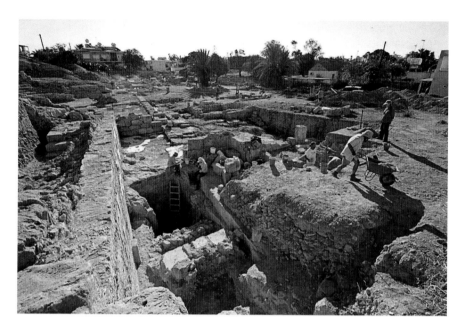

Fig. 5.2 Students participating in University of Sydney excavations of the Hellenistic-Roman theatre of Paphos, under the auspices of the Department of Antiquities of the Republic of Cyprus. Photograph by Bob Miller, courtesy of Craig Barker (Sydney University Museums)

has long been a key point of difference between sub-disciplinary areas at Sydney. Assessed fieldwork has never been compulsory to complete a major or Honors in Near Eastern or Classical Archaeology, and until 2010, no fieldwork courses were taught by staff who specialised in these areas. Until recently, Australianists, who were previously located in the "Prehistory" and "Historical Archaeology" sections, have taught all the field methods courses to undergraduates as part of the formal curriculum. Only Historical Archaeology and Prehistory ever ran assessed field schools before these became impractical in the 1990s. Students completing Honors in these programmes were previously expected to complete at least 5 days' fieldwork on an approved excavation, but Faculty-wide standardisation of Honors prerequisites forced cancellation of this requirement in the 2000s.

There are historical, geographical, practical and disciplinary reasons why Australianists have traditionally taught field methods to students as part of the curriculum, while staff in Near Eastern and Classical Archaeology have not. In the past, this difference has been more or less accepted as arising from the history and different cultures and traditions of the departments involved (see below). In a changed budgetary environment with a combined department, this difference increasingly raises issues about core vs. elective curriculum; the subject-specific knowledge students must learn to complete postgraduate research in the archaeology of different geographical regions and time periods; training graduates for employability in the local heritage industry; alignment between staff research interests and their teaching responsibilities; workloads and resources to support field and practical work teaching.

The University of Sydney does not currently teach any formally assessed field schools; field methods are taught by practical classes on campus (see below). Good students in all kinds of archaeology in the Department, and especially those doing research degrees, are normally given priority and support from research grants and bequest funds to attend university and other approved excavations and field schools in Australia and overseas. Senior students are expected to gain familiarity and experience with sites, artefacts and ways of doing archaeology in their chosen geographical area and chronological period, as is common elsewhere (e.g. Ucko et al. 2007). The University of Sydney is not unique in this respect; La Trobe University, University of Melbourne and University of Queensland also did not run assessable field schools in 2010–2011 (Table 5.2).

5.2.1   Origins and Growth (1948–1970s)

The first Department of Archaeology in Australia was founded at Sydney in 1948 with exclusive focus on the "Near East" and Ancient Mediterranean. Staff were appointed to the Department of Anthropology from the 1960s to develop Pacific and Australian Aboriginal "Prehistory". Judy Birmingham, previously a Near Eastern specialist, left the Department of Archaeology in the 1970s and established Historical Archaeology (Ireland and Casey 2006; Jack 2006; O'Hea 2000). New fieldwork opportunities in Australian archaeology were welcomed by Sydney students of Near Eastern and Classical Archaeology who wanted more excavation experience. Students already participated in Departmental fieldwork in Cyprus, Palestine and Greece, but opportunities were limited and expensive. When the Archaeology Department refused to support local fieldwork training, the student Archaeological Society ran their own excavations during university vacations. Birmingham co-directed and trained students on historical archaeology excavations at Irrawang in New South Wales (1967–1975), Wybelenna in Tasmania (1969, 1971) and many other projects that helped establish historical archaeology in Australia (Ireland and Casey 2006; Jack 2006).

Students also worked on Aboriginal sites. However, these generally involved digging small trenches with limited stratigraphical variation, few structures and no ceramics. Historical archaeology provided students with fieldwork experience considered more useful for participation on excavations in the Near East and Mediterranean, e.g. built structures; a variety of archaeological features, contexts, and stratigraphy; larger open-area excavations and a greater diversity of artefact types including ceramics. There was growing nationalistic and research interest in developing Australian archaeology to discover more about the past of an entire continent that was still largely unexplored by archaeologists (e.g. Murray and White 1981). This was before heritage legislation, and inexperienced students could excavate without a permit. There was no requirement to obtain permission from Aboriginal communities or others. There were no health and safety regulations. Time pressures were less important and students did not have to pay fees to attend university. Lectures could be cancelled at short notice, for example, so students and staff could run emergency rescue excavations at Town Hall Square in Sydney's city centre when building work threatened a colonial burial site in 1974 (Jack 2006: 23).

5.2.2 The Regentville Field School and Other Practical Classes (1980s to Late 1990s)

From 1985 until 1997, Historical Archaeology ran regular assessed field schools at the site of the Regentville colonial mansion (1823–1869) near Penrith in western Sydney. Students attended compulsory excavations and spend significant amounts of time on post-excavation analysis. Excavations were directed by Birmingham and Andrew Wilson under a NSW government permit issued to improve archaeological methods, teach students practical skills, and bring archaeology to the general public (Wilson 2000).

When Colley arrived in "Prehistory" in 1990, the section had seven academic teaching staff on permanent or long-term contracts plus tutors, dedicated technical and administrative support in the Department of Anthropology, exclusive use of a large four-wheel-drive vehicle for fieldwork and a healthy budget. Prehistory split from Anthropology and merged with Historical Archaeology in the mid 1990s. Colley team-taught practical courses and helped run regular field trips to archaeological sites in the Sydney area and beyond. Two or more staff co-teaching a class of 15–20 students was normal, and many separate in-depth practical courses were offered (e.g. field methods and surveying, animal bones, midden analysis, stone tools, rock art recording, historic ceramics, human skeletal remains, soil science). The department was allowed to limit enrolment numbers to match the availability of resources and equipment and to avoid overcrowding.

5.2.3 Funding Cuts and Major Change (Late 1990s to Mid-2000s)

Significant budget cuts started from the 1990s, and staff who left were not replaced. Student interest in Archaeology remained high, but income from enrolments no longer covered costs. The university banned ceilings on class sizes as it became necessary to enrol more students, and it was no longer economic for staff to co-teach practical classes simultaneously. By 2004, only 2.5 full-time academic staff remained in the "Prehistoric and Historical" programme; the vehicle and all tutors' positions were gone. Funding to maintain and update laboratory and field equipment dried up and facilities became outdated and run down. Entire areas of curriculum content were cancelled, including most specialist practical classes, and other course content become more generalised. Core curriculum was maintained by redesigning teaching, for example, by offering different units of study on a 2- or 3-year cycle rather than every year, and by accepting goodwill teaching help from research staff, associates, and local heritage organisations (e.g. the Australian Museum and the Sydney Harbour Foreshore Authority) that were concerned about the demise of the department.

From 1998 until 2002, Colley coordinated a new "Field/Laboratory Project" unit of study that replaced several previous specialist field methods and practical classes. This allowed senior students to gain degree credit for 35 h of assessed participation in practical work placements. Hours were based on university standardisation of degree structures linked to government funding and the introduction of student fees through the Higher Education Contribution Scheme (HECS). Each unit of study in a degree programme has a fixed and maximum number of teaching hours and notional assessment word length depending on its credit point weighting.

Students completing the "Field/Laboratory Project" were supervised on a good-will basis by over 40 archaeologists including university staff, postgraduate research students, and heritage professionals on research and consultancy projects in Australia and overseas. Enrolments rose from c. 25 to 40 within 4 years. Colley used the unit to conduct research into the knowledge and skills that heritage professionals expected students to learn so that they could work in archaeology, and how to assess students' learning (e.g. Colley 2003, 2007b). The unit was viable due to the high concentration of researchers and professional archaeologists working in the region or based in Sydney who had projects that could incorporate students. This is not the case everywhere in Australia and tightened OH&S and other regulations now present greater challenges. Flinders, La Trobe, UWA and UNE also teach or have previously taught field methods using a similar model (Table 5.2 and Wendy Beck, personal communication). At Sydney, growing enrolments created an increasingly unsustainable workload as more staff left the department without replacement and funding for administrative and technical support was also reduced. The administrative costs of running the "Field/Laboratory Project" unit became too high, and in 2004–2005, the unit was restricted to third year students completing an Archaeology major. Enrolments then dropped and it was later cancelled to comply with university financial policies.

5.2.4 Consolidation, Reconfiguration and New Approaches (Mid-2000s to 2010)

Teaching staff in Australasian archaeology at the University of Sydney increased slightly from the mid-2000s thanks to the departments' strong research profile, although most new appointments were on casual or short-term contracts. All remaining staff in Archaeology and Heritage Studies were eventually merged into a single organisational unit. Teaching of heritage studies, archaeological method and theory, and lecture-based content relevant to staff research areas were strengthened.

Fig. 5.3 University of Sydney students digging nineteenth century contexts at the Rocks, Sydney, for a 2006 training excavation directed by Helen Nicholson on behalf of the Sydney Harbour Foreshore Authority. Photograph Russell Workman

Hands-on teaching of archaeological computing and scientific methods continued through the University's Archaeological Computing Laboratory and Australian Centre for Microscopy and Microanalysis (Reade and Field 2008), respectively, and a new practical unit in stone tools analysis was developed and taught by postdoctoral researchers.

From 2006, a single generic "Field Methods" unit was offered to all senior Archaeology students. Short-term excavations of historic sites in Sydney's Rocks area by Sydney Harbour Foreshore Authority provided opportunity for students to gain local fieldwork experience as part of the course for the first couple of years (Fig. 5.3). A staff member was awarded a university teaching grant to build an on-campus simulated excavation box following the University of Queensland model (Hall et al. 2005). This initiative was stymied by university administrators who insisted the project be managed centrally to comply with OH&S policies. After some delay, the university quoted commercial rates to design and build an elaborate construction several times the original budget estimate. As no extra funding was available, the project was cancelled (Dougald O'Reilly, personal communication).

The revised Field Methods unit is taught on campus and introduces students to basic field survey techniques listed by Gojak (2007) as the minimum required to work on a field project. A new generic "Laboratory Methods" unit was introduced in 2007. So far these units have been offered in alternative years to all senior

Archaeology students and typically attract enrolments of over 75 each. Both units are co-delivered by 2–3 staff to manage student access to limited space and equipment. "Laboratory Methods" is subsidised by research staff and heritage consultants who donate access to their archaeological collections and supervise the students in return for their help in sorting artefacts. In 2009, Colley and Martin Gibbs obtained a university teaching grant to produce video clips of staff demonstrating techniques in field survey and lithics manufacture. These were made available to students on You Tube for 2010 and have also been used for teaching at University of New South Wales and UNE (Estelle Lazer and Alice Storey, personal communication). Colley and Gibbs (in preparation) are assessing University of Sydney student survey responses about the effectiveness of this teaching method.

Training students to perform atomised practical tasks (e.g. setting up survey gear, measuring and setting out a grid square, drawing a simple scaled map) is certainly useful. However, doing archaeological fieldwork involves applying such techniques

Fig. 5.4 University of Sydney students working with members of the local Aboriginal community on developer-driven excavations at Doonside, western Sydney, 2010. Photograph by Stirling Smith, courtesy of Comber Consultants Pty. Ltd.

in an appropriate way in an "authentic" context (cf. Perry 2004) to address research and other questions. This requires judgement, an understanding of relationships between practice and theory, and experience that can only be acquired in the field (Colley 2003, 2007b).

University of Sydney staff have forged research, professional and teaching collaborations with local researchers and practitioners through initiatives such as the Archaeology of Sydney Research Group (University of Sydney 2010a). Honorary associates, local consultancy companies and government heritage agencies regularly offer student access to collections and fieldwork experience and co-supervise Honors thesis research. Sydney students interested in studying Australian archaeology and heritage at Honors and postgraduate level volunteer on Australian and overseas archaeology projects whenever they can (Fig. 5.4). Depending on their competency, some students secure paid employment in the heritage consultancy industry which also helps pay university fees while studying. Involvement in such projects offers experiences that are as authentic and useful for possible future employment in the heritage industry as formally taught field schools. Field schools offer different kinds of learning experiences and are also very important. Running assessed field schools at University of Sydney is currently very difficult for funding, organisational and other factors that seem to be more or less significant at different Australian universities. This creates major variations in provision of field schools and fieldwork teaching nationally.

5.3 Factors Important to Field School Teaching in Australian Universities

5.3.1 University Funding, Restructuring and Local Organisational Cultures

Archaeology in most Australian universities is now located in "departments", "discipline areas" or "programmes" within larger jointly funded and managed multidisciplinary organisational units such as Schools or Colleges that have Arts and Humanities (A&H), Social Science and/or Science focus (Table 5.1). Other academic and administrative staff in the same organisational unit may be more or less understanding of the practicalities and costs of doing archaeology. The economic viability of discipline areas is subject to constant review by university management and there is internal and external competition for resources (e.g. staff and infrastructure) and income (e.g. research grants, sponsorship, donations, bequests and student fee income).

Running archaeological field and practical work teaching is comparatively costly. This is not recognised by government funding models that rank Archaeology with "lower cost" Arts and Humanities disciplines such as History, English and Philosophy that use cheaper teaching methods (e.g. lectures and discussion-based tutorials).

At universities where Archaeology can be taught within science or social science curriculae, a higher band of government funding applies (Table 5.1) though such courses may attract higher fees that impact on enrolments. Some universities allow departments to cap enrolments in courses that require access to limited technical equipment and space, while others do not. What works well in one university may be unviable, impossible or inappropriate elsewhere depending on local structures, policies, politics and people. Cultures of cooperation facilitate the running of field schools and practical work teaching, while cultures of competition create challenges.

The universities listed in Tables 5.1 and 5.2 offer general undergraduate degrees (e.g. Bachelor of Arts, Science, Social Science) with a major in Archaeology. Only some institutions (e.g. ANU, La Trobe and Flinders) also offer specialised undergraduate degrees in Archaeology. For example, a Bachelor of Archaeology is run cooperatively and jointly between Flinders University, University of Adelaide and University of South Australia with a capped quota of thirty students (Flinders University 2010). In this case, a culture of cooperation benefits Archaeology teaching. In other circumstances and institutional contexts where Archaeology has been unable to compete with larger, more powerful or "cost-effective" discipline areas, teaching programmes are threatened. Such factors contributed to the demise of Anthropology and Archaeology teaching at Charles Darwin University in the mid-2000s (Fredericksen and Walters 2002).

Some departments of Archaeology have more discipline-specific technical and administrative support (Fig. 5.1), access to better infrastructure (e.g. laboratory facilities, vehicles, fieldwork equipment, on-campus museum and reference collections) or are better placed to subsidise these from research or other income. Archaeologists frequently negotiate arrangements with other disciplines (e.g. in sciences) to access specialist equipment and with external organisations (e.g. museums, government heritage agencies) to access sites and collections to support their activities. For example, UNE runs some field schools in conjunction with staff in environmental sciences and central university support for intensive teaching modes, as part of distance learning in which UNE is a national leader, and facilitates archaeological field school teaching. UWA recently started running a 3-week field school out of semester, in collaboration with the Rio Tinto Iron Ore company's Cultural Heritage team in Western Australia's Burrup Peninsula. The region contains many important Indigenous Aboriginal rock art sites and cultural places. The field school will run for the next 7 years as part of Rio Tinto's Conservation Agreement linked to National Heritage Listing of the Burrup. Rio Tinto contributes money towards flights, food, and accommodation and numbers are capped at twelve students with participation assessed through an Arts Practicum unit of study. Students who cannot attend the field school are offered opportunities to develop the same skill set for the same duration of time through assessed participation in staff research projects (Alistair Paterson and Liam Brady, personal communication).

Increasingly complex regulation of OH&S and insurance for field projects, centralised university management processes, and "audit cultures" (Hamilakis 2004) challenge archaeological fieldwork practice. For example, the University of Sydney's online financial management system requires staff access fieldwork funds through corporate credit cards. As fieldwork often occurs in areas with limited or no electricity, let alone banks or automated banking facilities, this directive is impossible. University colleagues working abroad experienced difficulty obtaining financial approval for pigs bought without a receipt as gifts to local landowners as part of negotiating field access. In such circumstances, archaeologists may need to seek special consideration to accommodate standard disciplinary practice. Administrative work in managing field projects, especially those involving students, can be significant. This may not be recognised by generic workload formulae based on standard models of Arts and Humanities practice where such costs do not apply. Where archaeologists have more direct input into administrative systems and where managers and colleagues in other discipline areas understand archaeological practice, it is easier to run field schools and practical teaching. Such issues are not unique to Archaeology or to Australia (e.g. Aitchison 2004; Ucko et al. 2007). However, a recent review of Australian higher-education commissioned by the Labour government presented evidence for very high student to staff ratios and significant decline in teaching quality following years of under-funding by government. Australian higher-education providers overall were more heavily reliant on income from non-government sources and fee-paying overseas students than any other OECD country (Australian Government 2009).

5.3.2 Changing Circumstances and Attitudes of Students

Incorporating professional level training into undergraduate teaching is impractical at some universities for reasons of funding and lack of interest from a broad mass of students who now study archaeology for interest, not because they want to work in archaeology. Students commonly work to pay university fees, and living costs in major Australian cities are high. Even when they are interested, limited time and money prevents many students from participating in field schools and site visits at weekends or out of semester (e.g. Gibbs et al. 2005). Only some students are willing or able to volunteer on archaeological projects or can pay extra to participate in field schools. As universities regulate and standardise teaching contact hours and assessment, timetables are full and clashes must be avoided. It is hard to run a compulsory field school away from campus when students are only required to attend for 3 h per week and must return immediately to campus to attend other lectures. It may be possible to organise Archaeology teaching timetables to avoid clashes and accommodate field school attendance. However, most undergraduates are also studying non-Archaeology classes and some timetable overlaps are unavoidable. Field schools are most practical when taught in intensive 1–3 week blocks outside the normal teaching semester. Many students will now only participate in learning activities that are directly assessed towards degree completion. Fredericksen (2005) experienced such issues for field schools previously taught at Charles Darwin University, Northern Territory. Similar reasons prompted staff at University of Queensland to develop the TARDIS simulated archaeological site to teach field methods on campus from 1996 (Hall et al. 2005).

5.3.3   Research vs. Teaching

Fieldwork teaching has traditionally been aligned with research as part of a mentoring process. Participation in staff research projects allows students to access sites and collections and learn through supervised experience. Student participation helps staff complete their research. This model assumes manageable numbers of motivated and competent students, and adequate time. Limited resources, workload pressures, high student to staff ratios and large cohorts of disengaged students do not motivate staff to sacrifice their precious research time to undergraduate practical work teaching. It is much easier and cheaper to deliver lecture-based classes to the broad mass of students and restrict field and practical work teaching to research-capable senior and postgraduate students with genuine career interests in archaeology. This especially applies in research-focused universities including ANU, Sydney, Melbourne and University of Queensland that rank high in national and international research league tables. Due to current government funding models for research vs. teaching, in many universities the cost-effective production of high impact research outputs takes precedence over delivery of teaching to undergraduate students unless they pay fees that generate significant income.

5.3.4   Geography, Incidental Fees and Global Higher-Education Markets

The geographical focus of staff research and the location of fieldwork areas are relevant to field schools. Many Australian-based university researchers work in other countries and the Australian Research Council funds as many overseas archaeology projects as it does to Australian ones (Bowan and Ulm 2009). Public and student interest in archaeology overseas (e.g. Europe, Mediterranean, Egypt, South and Central America, West and South East Asia) influences the marketability of field schools in a globalised higher-education business. Although no data are available, it seems likely that more Australian students are willing to pay extra for field schools located overseas than in Australia. Several Australian universities run assessed overseas field schools, some of which cost students several thousand dollars to attend, including travel, food and accommodation (Table 5.2).

It is often as cheap to fly to neighbouring countries in south east Asia than to travel significant distances within Australia. Accessing remote Australian field areas from the major centres where most universities are located (Fig. 5.1) can be particularly expensive. The relative strength of the Australian dollar makes the cost of living cheaper in many south east Asian countries for Australians, and this impacts on choice of location for fieldwork and field schools. For example, Cooke (2008) as President of the Canberra Archaeological Society expressed concern about the demise of local Australian fieldwork training for students at ANU when the university's long-running field school at the historic mining settlement of Kiandra (Australian

Commonwealth Territory) was replaced by other field schools in the Pacific and South East Asia linked to staff research.

Flinders University successfully markets field schools to students from other countries willing to pay higher fees and travel costs for quintessentially Australian experiences in maritime, historical and Indigenous archaeology linked to staff research in South Australia and elsewhere. Many projects are run in interesting coastal, rural and remote locations and some involve working closely with Indigenous Aboriginal communities. Income generated is used to subsidise participation by Flinders students who pay smaller charges for local travel and accommodation (Flinders University 2010). Attracting international students as part of the "international education industry" is a key priority for the South Australian state government working in partnership with Adelaide City Council, universities and others. In 2007, international education was "South Australia's fourth biggest export earner generating $637 million in 2007 and supporting 3,250 jobs" with an estimated future capacity "to be a $2 billion export earner for South Australia by 2014" (Education Adelaide 2008: 2). Significant resources are put into building this market, and this business model presumably assists universities in South Australia to fund archaeological field schools.

Some students find extra money to participate in more expensive field schools, but this raises issues of equity and access for those unable to pay. The federal government regulates ancillary or incidental fees that universities may charge students for goods and services associated with degree completion (Australian Government 2005). A key principle is that Commonwealth-supported students and domestic fee-paying students *generally* must be able to complete the requirements of their course of study without the imposition of fees that are extra to HECS or tuition fees. Universities may charge students for some extra costs of running units of study if these are not compulsory for degree completion, or where students are offered a free equivalent unit. Universities develop their own incidental charges policies to comply with government regulations (e.g. University of Queensland 2010). These are open to interpretation and may vary between Faculties and other organisational units (e.g. University of Western Australia 2010). Analysis of university incidental fee policies is beyond the scope of the paper, but this is clearly an ambiguous area. For example, under central University of Sydney policies (University of Sydney 2010b), students can be charged incidental fees "essential for costs of food/transport/accommodation associated with field trips which are essential to the course". This contradicts current Faculty of Arts policies that apply to Archaeology. Charging undergraduate students hundreds of dollars or more for compulsory field schools is impractical as well as inequitable. Few students could afford to study Archaeology, enrolments would drop and the economic viability of the department would be threatened. Many University of Sydney undergraduates do pay to attend voluntary field schools and some receive travel scholarships from bequest funds and private foundations.

Charging incidental fees for compulsory field school participation is more realistic for specialised programmes such as the ANU's Bachelor of Archaeological Practice where students must complete one or more units of study that include fieldwork.

In some years, the only fieldwork unit offered is a field school that attracts incidental fees (Peter Hiscock, personal communication). Students have no alternative but to pay extra costs unless (presumably) they go part-time and defer degree completion until a cheaper unit becomes available. However, students serious about learning archaeology and field practice presumably already expect to have to pay extra for their education. Most Australian students already pay fees and other costs towards their university education themselves or through government loan schemes and do not expect everything to be free of charge.

5.4 The Past and Future of Archaeological Field Schools

Archaeologists working in the heritage industry, and members of the public interested in archaeology, sometimes contact the University of Sydney to ask staff to "run a field school" by involving students in proposed field projects. While the department actively encourages and supports student involvement in approved fieldwork on a voluntary extra-curricular basis, it is currently impractical for the department to run field schools. There seems to be some nostalgia about field schools as the "ideal" way to learn field practice and to initiate students into the profession. Many professionals learned field practice as students by participating in field schools, including those run at University of Sydney in the past. Some interested amateurs volunteered on archaeology projects run by Judy Birmingham and others up to the late 1990s. The BBC's "Time Team" television show is also very popular in Australia and has encouraged public interest in geophysical survey and "going on a dig" which is less practical in Australia than the UK.

While discussing field training in Australia as part of research for this paper, some archaeologists volunteered information about their own field school experiences as students. More stories concerned people, places, travel and adventures than the specifics of learning archaeological practice. Experiences were both positive and negative and such aspects of student learning on field schools have been discussed elsewhere (e.g. Fredericksen 2005; Perry 2004). Reactions of teaching staff to field schools were mixed; some were very positive and enthusiastic, while others emphasised significant challenges and difficulties.

There are many reasons why some Australian universities teach many field schools, some teach a few, and others teach none at all. Professional archaeologists complain that Australian universities do not teach students enough Australian archaeology, cultural heritage management, or fieldwork methods essential for local heritage industry employment (Gojak 2007). Running assessed field schools is an important way for students to learn these skills, but only some universities are currently willing or able to afford the costs of such teaching. It is technically possible for a student from a university with no field schools to enrol in field school units at another university for degree credit, and some students do. However, due to travel distances between Australian universities (Fig. 5.1), this is not always practical or cost-effective.

Without accreditation, the archaeology and heritage profession has no formal responsibility for graduate training. Voluntary joint initiatives in education, training and employment between professionals and universities have been brokered by national associations including the Australian Archaeological Association, the Australian Institute for Maritime Archaeology and the Australian Association of Consulting Archaeologists. These include a national work experience register for students and employers and a project to benchmark Archaeology Honors degrees (Australian Archaeological Association 2010b; Beck 2008). One possible solution to major differences in field training opportunities nationally would be for professional archaeological organisations, state heritage agencies, and private consultancy companies to take joint responsibility for running and subsidising the costs of formally taught and assessed field schools across Australia in collaboration with a national consortium of university departments. Students could be asked to make some financial contribution and scholarships could be provided for excellent students and for those in financial difficulty. Such an initiative would require the establishment of a national body to speak for all aspects of Australian archaeology, as has already been proposed in other contexts (Gibbs et al. 2005). Until this happens, the provision of field school teaching is likely to remain variable and patchy between Australian universities for some time to come.

Acknowledgements I would like to thank people who helped with research for this paper. Martin Gibbs commented on an early draft, Annika Korsgaard provided research assistance and created Fig. 5.1, and the following clarified information on university websites or provided useful insights: Wendy Beck, Alison Betts, Liam Brady, Peter Hiscock, Dougald O'Reilly, Estelle Lazer, Helen Nicholson, Alistair Paterson, Dan Potts, Ted Robinson and Matthew Spriggs.

References

Aitchison, K. (2004). Supply, demand and a failure of understanding: addressing the culture clash between archaeologists' expectations for training and employment in "academia" versus "practice". *World Archaeology, 36*(2), 203–219.

Australian Archaeological Association. (2010a). How do I study to become an archaeologist? Electronic document. Retrieved September 28, 2010, from http://www.australianarchaeologicalassociation.com.au/study_options

Australian Archaeological Association. (2010b). The register of archaeology work experience partners. Electronic document. Retrieved September 28, 2010, from http://www.australianarchaeologicalassociation.com.au/work_experience

Australian Government. (2005). Attorney-General's Department, Higher education provider guidelines (05/09/2005). Electronic document. Retrieved November 19, 2010, from http://www.comlaw.gov.au/comlaw/management.nsf/lookupindexpagesbyid/IP200510176?OpenDocument

Australian Government. (2009). Department of Education, Employment and Workplace Relations, Review of Australian Higher Education. Electronic document. Retrieved September 28, 2010, from http://www.deewr.gov.au/HigherEducation/Review/Pages/default.aspx

Australian Institute for Maritime Archaeology. (2010). AIMA/NAS maritime archaeology training course. Electronic document. Retrieved October 3, 2010, from http://aima.iinet.net.au/nas/aimanas.html

Balme, J., & Wilson, M. (2004). Perceptions of archaeology in Australia amongst educated young Australians. *Australian Archaeology, 58*, 19–24.

Beck, W. (2008). *By degrees: Benchmarking archaeology degrees in Australian universities.* Armidale, New South Wales: Teaching and Learning Centre, University of New England.

Beck, W., & Balme, J. (2005). Benchmarking for archaeology Honours degrees in Australian universities. *Australian Archaeology, 61*, 32–40.

Bowan, J. K., & Ulm, S. (2009). Grants, gender and glass ceilings? An analysis of ARC-funded archaeology projects. *Australian Archaeology, 68*, 31–36.

Colley, S. (2002). *Uncovering Australia: Archaeology, indigenous people and the public.* Washington: Smithsonian Institution Press.

Colley, S. (2003). Lessons for the profession: Teaching archaeological practical work skills to university students. *Australian Archaeology, 57*, 90–97.

Colley, S. (2004). University-based archaeology teaching and learning and professionalism in Australia. *World Archaeology, 36*(2), 189–202.

Colley, S. (2007a). Public benefits of archaeology: Results from a student questionnaire. *Australian Archaeology, 65*, 30–36.

Colley, S. (2007b). Assessment of archaeological skills: Implications for theory and practice. In P. J. Ucko, Q. Ling, & J. Hubert (Eds.), *From concepts of the past to practical strategies. The teaching of archaeological field techniques* (pp. 159–168). London: Saffron.

Colley, S. & Gibbs, M. (in press). Capturing archaeological performance on digital video: Implications for learning archaeology. *Research in Archaeological Education.*

Colley, S., Ulm, S., & Pate, F. D. (Eds.). (2005). *Teaching, learning and Australian archaeology. Australian archaeology* (Vol. 61). Australian Archaeological Association Inc., Canberra, Australian Capital Territory.

Cooke, H. (2008). Canberra Archaeological Society, 2008 President's report. Electronic document. Retrieved November 20, 2010, from http://www.cas.asn.au/about.php

Coutts, P., & Wesson, J. (1980). Victoria archaeological survey summer schools in archaeology: An evaluation. *Australian Archaeology, 11*, 119–127.

Education Adelaide. (2008). South Australia's International Education Industry. Submission for the Bradley Committee by Education Adelaide on behalf of its stakeholders. Electronic document. Retrieved September 28, 2010, from http://www.studyadelaide.com/library/Submission%20to%20Review%20of%20Higher%20Education.pdf

Eslick, C., & Frankel, D. (2006). Judy in the sixties: An inspiration. *Australasian Historical Archaeology, 24*, 17–18.

Feary, S. (1994). Teaching and research in archaeology: Some statistics. *Australian Archaeology, 39*, 130–132.

Flinders University. (2010). Welcome to archaeology. Electronic document. Retrieved November 19, 2010, from http://www.flinders.edu.au/ehlt/archaeology

Frankel, D. (1980). Education and training in prehistory and archaeology in Australia. *Australian Archaeology, 11*, 69–184.

Fredericksen, C. (2005). Archaeology out of the classroom: some observations from the Fannie Bay Gaol field school, Darwin. *Australian Archaeology, 61*, 41–47.

Fredericksen, C., & Walters, I. (2002). Archaeology from the Frontier: The past, present and future of research at the Northern Territory University. *Australian Archaeology, 55*, 30–34.

Gibbs, M., Roe, D., & Gojak, D. (2005). Useless graduates?: Why do we all think that something has gone wrong with Australian archaeology training? *Australian Archaeology, 61*, 24–31.

Gojak, D. (2007). The "Gojak list"; or Denis Gojak's tips for evaluating your performance as a cultural heritage practitioner'. In C. Smith & H. Burke (Eds.), *Digging it up down under. A practical guide to doing archaeology in Australia* (pp. 15–17). New York: Springer.

Hall, J. (1980). Archaeology at the University of Queensland. *Australian Archaeology, 10*, 79–85.

Hall, J., O'Connor, S., Prangnell, J., & Smith, T. (2005). Teaching archaeological excavation at the University of Queensland: Eight years inside TARDIS. *Australian Archaeology, 61*, 48–55.

Hamilakis, Y. (2004). Archaeology and the politics of pedagogy. *World Archaeology, 36*(2), 287–309.

Ireland, T., & Casey, M. (2006). Judy Birmingham in conversation. *Australasian Historical Archaeology, 24*, 7–16.

Jack, I. (2006). Historical archaeology, heritage and the University of Sydney. *Australasian Historical Archaeology, 24*, 19–24.

Megaw, V. (2000). Confessions of a wild colonial boy. Rhys Jones in conversation with Vincent Megaw. *Australian Archaeology, 50*, 12–26.

Murray, T., & White, J. P. (1981). Cambridge in the bush? Archaeology in Australia and New Guinea. *World Archaeology, 13*(2), 255–263.

O'Hea, M. (2000). The archaeology of somewhere-else: A brief survey of classical and near eastern archaeology in Australia. *Australian Archaeology, 50*, 75–80.

Perry, J. E. (2004). Authentic learning in field schools: preparing future members of the archaeological community. *World Archaeology, 36*(2), 236–260.

Reade, W., & Field, J. (2008). Putting the past under the microscope. In K. R. Ratinac (Ed.), *50 great moments: Celebrating the golden jubilee of the University of Sydney's electron microscope unit* (pp. 311–318). Sydney: Sydney University Press.

Smith, C., & Burke, H. (2007). *Digging it up down under. A practical guide to doing archaeology in Australia.* New York: Springer.

Ucko, P. J., Ling, Q., & Hubert, J. (Eds.). (2007). *From concepts of the past to practical strategies. The teaching of archaeological field techniques.* London: Saffron.

Ulm, S., Nichols, S., & Dalley, C. (2005). Mapping the shape of contemporary Australian archaeology: Implications for archaeology teaching and learning. *Australian Archaeology, 61*, 11–23.

University of Queensland. (2010). Handbook of University policies and procedures, 3.10.9 incidental and ancillary fees levied on students. Electronic document. Retrieved November 19, 2010, from http://www.uq.edu.au/hupp/index.html?page=25083

University of Sydney. (2010a). Department of Archaeology, The Archaeology of Sydney Research Group background. Electronic document. Retrieved November 19, 2010, from http://sydney. edu.au/arts/archaeology/research/archaeology_of_sydney_research_group/

University of Sydney. (2010b). Strategic planning office, courses and fees toolkit – Glossary. Electronic document. Retrieved November 19, 2010, from http://www.planning.usyd.edu.au/ courses_fees/glossary.php

University of Western Australia. (2010). Teaching and learning, incidental fees and charges. Electronic document. Retrieved November 19, 2010, from http://www.teachingandlearning. uwa.edu.au/students/fees

Wilson, A. (2000). Historical archaeological excavations at Regentville. Electronic document. Retrieved September 30, 2010, from http://sydney.edu.au/arts/archaeology/regentville/

Chapter 6
The UCLA Archaeology Field Schools Program: Global Reach, Local Focus

Ran Boytner

6.1 Introduction

The traditional North American archaeology field school is designed to train the next generation of professional archaeologists, whether in academia or CRM. Universities support such field schools in two ways. They allow faculty members to count teaching field schools as part of their normal teaching load and also provide financial assistance to field schools on top of funds collected through student tuition. Recently, a number of archaeologists have begun to write about field schools, with discussions focused on pedagogy and attempts to analyze the very scarce data publically available about the phenomenon of field schools in North American academic institutions (Baxter 2009; Perry 2004, 2006; Piscitelli and Duwe 2007; VanderVeen and Repczynski 2010). But the traditional North American field school concept is threatened by both declines in funding and the increased complexity of archaeological research (see Chap. 5, for similar constraints in Australia).

The impact of the current economic crisis is fast changing the capacity of universities to support archaeological field schools. The low student to faculty ratio together with the costs of field room and board, vehicles, and salaries make little economic sense to Deans and Provosts when attempting to fund on-campus classes with severely declining available funds. At the same time, the number of students who continue to pursue advanced degrees in archaeology after a field school experience remains small. Most field school directors see that as a positive outcome where the forces of natural selection weed out those not sufficiently committed to the rigors of the discipline and to a successful career in it, in contrast to the perceived benefits argued by UK academics even for those not continuing (see Chaps. 1 and 3). Students who do not continue and pursue a career in archaeology are an acceptable loss – a

R. Boytner (✉)
Institute for Field Research, Department of Anthropology, 120 Grace Ford Salvatori Hall,
University Park Campus, University of Southern California, Los Angeles, CA 90089, USA
e-mail: rboytner@gmail.com

H. Mytum (ed.), *Global Perspectives on Archaeological Field Schools:*
Constructions of Knowledge and Experience, DOI 10.1007/978-1-4614-0433-0_6,
© Springer Science+Business Media, LLC 2012

dead weight – so focus is maintained on those who do embrace the discipline. But while archaeologists are accepting the low continuation ratios, university administrators question the effectiveness of this investment and are shifting scarce resources to classes with much higher faculty to student ratios and where students may get less specialized education.

Universities are cutting back on field school sponsorship at the time when the archaeological endeavor is increasingly involving more analytical work and thus increasing costs of research. The discipline has experienced a shift in the past few decades from projects run by a single director to a more hard science model where teams of specialists often come together and gather ever larger quantities of data in increasingly sophisticated methods. A wide range of scientific techniques and analytical instruments is frequently deployed, and while these enhance the archaeologist's ability to interpret the material record, they also increase costs. Just as archaeologists need more resources to conduct advanced research, funding resources are shrinking.

One of the unanticipated consequences of the increased scientific nature of North American archaeological research is that results become less accessible to the public. Many archaeologists are talking in language that the lay public find hard to follow. In 2006, Brian Fagen wrote that the discipline may come to the "end of its Golden Age" (Fagan 2006:59). Archaeologists are losing their hold on the North American public imagination (unlike in the UK where interest and awareness continue to increase) and there is some danger that the discipline will go the way geography did half a century ago. While vibrant to its practitioners, it is rarely discussed in a public forum and it is usually conducted outside the public gaze. Geography now rarely figures explicitly on the pages of National Geography Magazine, and geographers are absent from the Discovery Channel.

At the turn of the millennium, the CIoA at UCLA attempted to rethink the place of archaeology in the twenty-first century, focusing on the future needs of the discipline. Among the many initiatives that emerged was the creation of an overarching CIoA Archaeology Field School Program (CIoA FSP). Launched in 2007 with a single field school and 11 students, by 2010 the program grew to 26 field schools and 280 students. At the peak of the program, there were UCLA archaeology field schools in every continent but Australia, and in 17 different countries, involving participation by many academics and universities as well as other agencies. While each university creates a unique environment for their field schools to operate and grow, the general outline and focus of CIoA FSP was simple and may be duplicated elsewhere.

It is important to note that the CIoA Field Schools Program did not limit itself to the financing of archaeological research or the training of the next generation of archaeologists. An important goal was to reach out to the many students who will not pursue a career in the discipline to create an experience that will encourage lifelong relationships with archaeology. The extensive outreach of the program has built new, educated, and enthusiastic audiences that will support the discipline for years to come.

To date, only limited research focused on modern archaeology field schools (Baxter 2009:17; also see Perry 2004, 2006). Field schools are usually not mentioned in archaeological literature that discuss, examine, and analyze archaeological curriculum or educational goals (Baxter 2009:18). In many cases, field schools are run as independent operations by North American university faculty members who develop individual field schools based on personal experience and with little standardization of methods, pedagogy, or logistical frameworks. Particular field schools seem to reflect the personality of the instructor and the culture of particular institutions (Baxter 2009:17). It has therefore been difficult to examine the field school institution beyond generalities. This study is an attempt to begin to address this *lacuna*. Using data collected from CIoA FSP activities over the past 4 years, 61 different field schools with 662 student participants can be analyzed. Although the CIoA FSP no longer exists, it was a large program that reached students from dozens of universities across the US and the world. The data presented here may therefore be used as a baseline for future research on field schools and can provide valuable insights into the present and future of the field school as a pedagogic activity.

6.2 Audience

Most archaeologists view field schools as training grounds for future professionals (see Gifford and Morris 1985; Haury 1989; Joiner 1992; Walker and Saitta 2002; also Chap. 3). They are not only places where students learn the first principles of archaeological field research, they are also the locus where individuals experience the rigors and demands of field work. It is the first trial where the romantic classroom notion of archaeology is confronted with the realities of field life (Baxter 2009:11; Perry 2006:26; Walker and Saitta 2002:199).

Despite the central role of field schools in the training of future generations of practitioners, most archaeologists recognize that the role of field schools is larger then that. Many students come because they desire adventure at home or abroad. They seek discoveries that are both intellectual and personal. They wish to learn about the past, but they also want an environment where they will be challenged physically and emotionally. Although they may not pursue the discipline any further, these students will continue to have an interest in the past many years after the successful completion of a program if their field school provided a positive and holistic educational experience (VanderVeen and Repczynski 2010:26; also see Baxter 2009; McManamon 1991; Perry 2004, 2006; Pyburn 2003; Staniforth 2008; Walker and Saitta 2002).

Longitudinal studies of the impact that field schools have had on student participants do not presently exist. It is unclear what is the proportion of field school participants that actually pursue careers in either academic or CRM archaeology. But anecdotal evidence suggest two important trends: almost all students find archaeological field schools a life-changing event, and many of current graduate students decided to pursue careers in archaeology after participation in a field school. Given that CIoA FSP was created only 4 years ago, it may not yet provide reliable data on these issues either, but it will act as a substantial sample against which general impressions can be set.

Table 6.1 Program-wide averages of students' response to 2009 and 2010 CIoA-FSP evaluation survey (2009, $N = 157$; 2010, $N = 235$)

	2009	2010
Program provided good training in archaeology	87.6%	82.5%
You have learned something which you consider valuable	94.4%	93.5%
Subject interest before the program	80.7%	90.7%
Subject interest after program	89.3%	93.8%

Standardized student evaluations for CIoA FSP field schools were only introduced in 2009. Prior to this, field school directors conducted their own individual evaluation processes and so comparison of results is difficult, as questions asked of students varied between programs. From 2009, CIoA FSP evaluation forms were standardized and it has become possible to conduct comparative evaluations of student experiences between field schools and across the whole program, with data now available for 2 years.

The CIoA FSP evaluations are both metric – asking students to provide numeric evaluations of specific elements of the field school – and expressive, where students are asked to answer questions in the narrative. Consistently across the board, many students write that the field school was a unique experience that made them understand the world in a new and different way (see also Chap. 7). Students are impressed by the exposure to cultures other then their own – whether ancient or in the locale where the field school takes place. But students are also impressed by what they learn about themselves. For most, it is the first time that they have traveled abroad, lived in small groups, and engaged in intensive physical labor. The exposure to primary research and the struggles over understanding the data or designing solutions to address real research problems emerging from field investigation also impress the students and allow an enhanced appreciation for scientific work.

The CIoA FSP evaluations show that field schools fulfill most student expectations and demonstrate how effective field schools are in providing positive learning experiences. Students gain an enhanced appreciation for the discipline and its goals by completing a field school (Table 6.1). There is no doubt that field schools are also places where individuals make the decision to pursue a career in archaeology. Most of the incoming archaeology graduate students to UCLA in the past 5 years participated in an organized field school during their undergraduate career. Many report that this was the event that either turned them to archaeology or solidified their commitment to a career in the discipline.

Despite the lack of systematic survey and long-term research of the issue, there is little doubt that the vast majority of students participating in archaeology field schools will develop careers outside the discipline. Perry suggested that in 2006 alone, over 1,400 US-based students participated in archaeology field schools (Perry 2006:25). These numbers are far too high to suggest all, or even most, pursued careers in archaeology (see Chap. 3, for UK data).

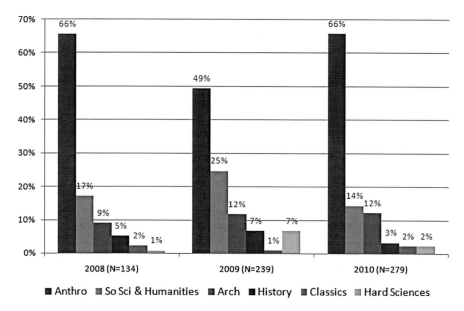

Fig. 6.1 Major distribution of CIoA field schools program students in the past 3 years

Data collected by CIoA FSP further support this assumption (Fig. 6.1). Only a very small proportion of the students identified archaeology as their major. It is assumed here that only students committed to the discipline will provide such identification for two reasons. First, most students are majoring in anthropology, the typical home for archaeological studies in the US as archaeology is a subdiscipline of anthropology. Anthropology students are keenly aware of the discipline and yet choose not to self-identify as archaeologists. Second, evaluations are filled at the last day of the field school, when students have a very strong sense what archaeology is and yet they still choose not to self-identify as member of the "guild."

6.3   Benefits

It is unclear why students chose to attend archaeology field schools as no systematic survey has been undertaken of participants on this issue. The sense of adventure VanderVeen and Repczynski wrote about is certainly an incentive (VanderVeen and Repczynski 2010:26). But the resources required for attending an average field school and the sheer numbers of students doing so suggest other forces are at work as well.

On June 10, 2009, the US House of Representatives approved the Senator Paul Simon Study Abroad Foundation Act (NAFSA n.d.). It was part of the Foreign Relations Authorization Act for Fiscal Years 2010 and 2011 (H.R. 2410). The

Simon Act sets the goal that, in 10 years, the number of US college students studying abroad should increase fourfold from the present 250,000 to 1 million annually.
To achieve this goal, the legislation establishes an innovative new structure that will
provide financial support to students for study abroad program. At the same time, it
encourages US higher education institutions to address the on-campus factors that
currently impede students' ability to study abroad. The legislation authorizes $40
million in funding for fiscal year 2010 and $80 million for fiscal year 2011. This
legislation has a large bi-partisan support both in the House and the Senate and is
viewed as part of the national security interests of the US. Indeed, the 9/11
Commission chairs endorsed the legislation and many other organizations see it as
an important component in advancing diplomatic and economic national goals. On
June 9, 2010, over 40 different organizations sent a letter in support of the legislation and wrote:

> Today's global demands challenge every sector of our economy, both public and private,
> and affect workers in each of our communities across the country. Our nation's economic
> competitiveness, diplomatic strategies and security efforts continue to rely on our ability to
> understand and communicate with the rest of the world. However, even with these increased
> demands for global skills, only about 1 percent of our students have the opportunity to study
> abroad each year (American Council on Education 2009).

The Simon Act created strong incentives for universities to increase the number
of students they send to international programs. Archaeology field schools usually
offer appropriate environments to teach students about lives other than their own.
They frequently involve remote locations; they include intensive studies of cultures
that are vastly different from the life of a typical US college student; they present
students not only with intellectual but also physical challenges (the physical challenge should not be underestimated for a society developing to be a service economy when most work is done in offices and in front of computers); and they are
intense learning experiences. They are seen as safe environments with low faculty
to student ratios that provide excellent study abroad opportunities to students who
may not wish to participate in a semester- or year-long program.

For students, archaeology field schools offer three important incentives: the
exotic factor, credits and costs, and access to faculty. Due to the high exposure of
archaeology in the public's imagination (from the *Indiana Jones* movies to *Time
Detectives* TV series), archaeology is viewed as an exotic discipline. While some
disappointment is frequently part of a student's actual engagement with the reality
of field work, discoveries are still exciting and most students are delighted with
finding objects from the past, although most are not real treasures.

Perry wrote that on average, archaeology field schools offer six semester credit
units at the costs of $1,000–6,000 per program (Perry 2006:25). To students out-
side North America, these costs seem prohibitive, but universities and the public
across North America view higher education as commodity and a path to higher
income levels. University education is thus priced accordingly; some universities
charge up to $50,000 in annual tuition and the costs per credit unit are much higher
than those associated with field schools. There is a broad agreement among academic
archaeologists in the US that field school cost should not exceed $5,000 and most

field schools are priced accordingly. Tuition above this level is seen as elitist, limiting the number and economic character of participating student population.

Another important incentive to students is that, on archaeology field schools, they gain unequaled access to faculty members. Because student to staff ratios in all field schools are low, and because staff live in the field with students for the duration of the program, students get to know faculty well. More significantly, faculty get to know students individually and can therefore write intimate, detailed, and insightful recommendation letters. In a US academic culture, a good recommendation letter carries significant weight. Students attending the normally crowded US universities rarely get to develop close relationships with their faculty, and archaeology field schools offer a unique opportunity for students. This is valuable not only to those wishing to develop a career in archaeology; given the competition to enter US graduate schools in medicine, law, or business, a good recommendation letter from a faculty that is outside the discipline which the student wishes to study can have a significant and positive impact in the evaluation of her/his application.

6.4 The UCLA Archaeology Field Program

The CIoA developed a number of initiatives to address new challenges to the discipline, of which CIoA FSP was one. The CIoA FSP mission statement reflects both a commitment for rigorous training of the next generation of archaeologists and a strong emphasis on using archaeology as a means to understand other cultures.

> The CIoA Field Program strives to be an educational leader in archaeology for the 21st century. Through programs covering the range of archaeological practice, students learn and experience the discipline through direct engagement with research projects directed by leading scholars in the field. Archaeology can be a transforming force and we seek to inspire. We wish to be the primary training grounds for the next generation of archaeologists. We also aim at nurturing life long relationships with archaeology for students who will pursue other careers. We believe that increased diversity is not just a desired goal but a daily practice in which both research and interpretations will become richer and global.

Moreover, CIoA FSP was not limited solely to the traditional archaeological endeavors – survey and excavation. It expanded that definition to cover the range of archaeological experience, and the participating field schools may be one of the following six types:

1. Field Archaeology programs focus on hands-on, traditional practices of archaeology. They include survey and/or excavation and are designed to allow students to have thorough training in field archaeology.
2. Ethnoarchaeology programs concentrated on analogies and the study of contemporary material culture among living communities. Their goal was to explore how contemporary objects and production processes may have been used in the past when observing similar objects in the archaeological record. In some cases, these programs used archaeological methods for the study of modern cultures that are difficult to study otherwise (Buchli and Lucas 2001; David and Kramer

2006). For example, the Arizona Migrant program studied material remains left by undocumented migrants crossing the Sonora Desert of Southern Arizona on their way from Mexico to the US (De León 2010). Insights gained from this research explore issues of migrant demographics, abuse, and adaptation to human and geographical landscapes.

3. Museum programs are designed to familiarize students with the life of artifacts after excavation and their presentation to the public. Participants were exposed to the technical, cultural, financial, and political dimensions of museum exhibition and strived to understand the complexities involved in museum work. Each program created as a final project an exhibit, usually of small scale and typically a single exhibit case.

4. Conservation programs introduced students to conservation practices of archaeological objects. The conservation programs were usually run in conjuncture with field archaeology programs, exposing students to conservation practices immediately from discovery, through lifting, final stabilization of objects, and transfer either to long-term storage or museum displays.

5. Science and archaeology programs emphasized analytical work within archaeology. While some of the field schools may be laboratory-based, most conduct science in the field. They included programs in geomatics, geophysics, paleoethnobotany, zooarchaeology, and use of hard sciences in the archaeological endeavor.

6. Travel/Study programs operated in those countries which do not legally allow archaeological field schools to be held in their territories. In these rare cases, field schools did not engage with field archaeology, but instead traveled extensively and studied intensively the cultural heritage of the host country on site.

All CIoA FSP field schools were attached to research projects where student experienced – and took part in – actual archaeological research; none were purely teaching and learning exercises. While the research methods, location, time periods, and theoretical approaches varied greatly between the field schools, all were data-oriented research projects that were associated with academic inquiry. There was a strong emphasis on anthropological archaeology, but this approach was not mandatory and students were exposed to other traditions of archaeological emphasis and interpretation.

Each of the CIoA FSP field schools offered 12 credit units, which is the number of units students at UCLA take in a full term. UCLA is a Quarter-based education institution, in which each term is 10 weeks. Some other US universities are using the Semester system (13–16 weeks each term). The typical conversation of credit units between quarter to semester-based systems is a ratio of 3:2, and UCLA 12 credits usually translate to 9 semester units. While these calculations may seem trivial, they carry great importance for students. Many students can participate in archaeology field schools only with the assistance of Financial Aid schemes (see Perry 2006:25). Students may claim a full quarter Financial Aid package if they receive at least 12 quarter or 8 semester units during the summer. It is because of this reason that the high number of credits awarded by CIoA FSP allowed students better access to financial resources and made the program popular.

The increase in average field school costs (Table 6.2) and the challenging economic conditions of the past 3 years had a direct impact on student financial needs.

Table 6.2 Breakdown of CIoA FSP student numbers participating in individual field schools

	2007	2008	2009	2010
Number of CIoA programs	1	13	21	26
Number of students attending CIoA programs	11	134	237	276
Number of Universities from which students came on CIoA programs	3	53	88	111
Average of tuition fees on CIoA programs UC students in US $	$3,700	$3,819	$4,108	$4,777
Average of tuition fees on CIoA programs non-UC students in US $	$4,000	$4,119	$4,408	$5,174
Percentage of students obtaining financial aid for field school	N/A	33%	34%	38%
Percentage of female students on CIoA programs	82%	75%	67%	70%
Percentage of students self-identifying as archaeology majors	N/A	9%	12%	12%

The portion of students needing assistance continues to grow and over a third of all CIoA FSP participants received some type of assistance.

The CIoA Field Schools Program grew rapidly in its 4 years of existence. From a humble beginning with a single field school in 2007, the program has evolved by 2010 to include 26 different field schools and over 270 students (Table 6.2). The average field school had 11 students, but some had as few as three and others as many as 28 students. The CIoA FSP did not limit the number of students that could participate in any program, but rather restricted the ratio between staff and students and did not allow a ratio higher than 1:7. In most programs, the ratio was usually one staff member to each three students, although some had a 1:1 staff to student ratio.

Each CIoA FSP field school was attached to an ongoing research project directed by a faculty member and has graduate students and other professionals serving as staff members. The scale of CIoA FSP far exceeds the number of available UCLA faculty, and many of the field schools were directed by archaeologists from peer research institutions. Peer scholars entered into partnership relationships with CIoA FSP and run the field schools as joint research projects (see more about the tensions such arrangement brought in the Epilog).

Two of the most important partners were the Institute of Archaeology at University College London (UCL IoA) and the Center for Advanced Spatial Technology (CAST) at the University of Arkansas, Fayetteville. The partnership with CAST was established due to their global leading position in geomatics, the collection and study of spatially referenced data. As most future archaeological research will involve some type of geospatial analysis – whether site-, regional-, or analytical-based investigation – student exposure to geomatics capabilities is highly desirable. Given that many of the participating students will not pursue a career in archaeology, geomatics may be the only useful technical tool they may take with them as the methods, theory, and practice of geomatics are used across the board in almost all contemporary human endeavors (among many, see Gomarasca 2009; Harrower 2010; Hodder 2005; Kavanagh 2003; Li et al. 2007; Wolf and Ghilani 2006; but not Bender et al. 2007; Tilley 1994; Tilley and Bennett 2007).

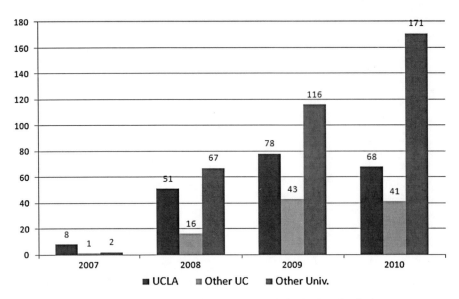

Fig. 6.2 Student enrollment on CIoA field schools, by home institution distribution

CIoA FSP was designed to attract students beyond the UCLA campus and to be inclusive by admitting students regardless of their major of study. The first CIoA FSP field school attracted students from three different institutions (UCLA, University of Chicago, and UC Riverside). By 2010, CIoA FSP attracted students from 111 different universities (Table 6.2), mostly from US-based institutions but including students from across the world (Canada, New Zealand, Israel, France, Australia). This student diversity was generated, in part, from the diversity of the faculty and their home institutions who participated in the program. It also reflected the increasing reach of the program to universities not traditionally training their students in archaeology field work.

The 2008–2009 economic crisis experienced throughout the world affected California particularly severely, and this can be seen in enrollment numbers. Between 2007 and 2009, student enrollment from UCLA and other University of California (UC) campuses increased each year, but the numbers fell in 2010, with a decline of both UCLA and UC student enrollment (Fig. 6.2). Increasing numbers of students in states that suffered most in the recession were unable to afford the additional costs of summer classes and frequently had to spend the summer working to save funds to pay for education in the following year (for similar observation, see Perry 2006:25). The impact was particularly hard felt among UCLA students, as the UC Regents approved a steep, 32% tuition increase on November 18, 2009 to add to their financial commitments.

Efforts to recruit students from diverse academic departments yielded mixed results (Fig. 6.1). Students claiming to be archaeology major were consistently a small minority, representing approximately 10% of students attending field schools.

In the US, archaeology – with Boston University as the sole exception – is not an independent department. Instead, undergraduate level archaeology may be taken as a subfield specialization in many departments (Classics, Near Eastern Languages and Cultures, or Religion), but usually it is taught as part of Anthropology. Self-identification as archaeology major therefore implies a strong commitment to the discipline. At the same time, the vast majority of students who attended CIoA FSP field schools were anthropology majors (whether BA or BS), indicating that the program had a strong attraction to its core audience in anthropology.

Significant numbers of students from other Social Science or Humanities majors also found attending the CIoA FSP field schools attractive. Students coming from History and Classics usually consisted of at least a quarter of all attendees. Study abroad programs are usually within the Social Sciences or the Humanities, so these patterns are consistent with those observed among all US students attending study abroad programs (see 2009 Open Doors Report by the Institute of International Education: Table 22).

Many North American archaeologists see themselves as the most "scientific" of all the social sciences, yet the data show that despite strong emphasis of scientific research and analytical methods, CIoA FSP did not attract many students from the hard sciences. These students are usually seeking internships in their chosen major during the summer, and attending archaeology field schools is seen as an indulgence and counter-productive to their careers, a distraction viewed negatively by both their peers and their advising faculty. This is particularly true for leading research universities (R1 institutions), where students are inducted into seeking advanced degrees in their specialization. Archaeology is not seen as a means to advance an application to a respectable graduate school in the hard sciences, although there is a strong exception for students seeking to apply for medical school. For these students, an archaeology field school can be seen as thinking "outside the box" and give them a slight edge of the very competitive application process. Medical schools require standard preparation, but are struggling to distinguish between numerous candidates with similar qualifications and experience. An archaeology field school experience can stand out as unusual, but still within the rigors of scientific research, and so can help potential medical school candidates. The lack of students from hard science majors, however, is disappointing. This is certainly an area where archaeology field schools generally are challenged. More targeted recruiting is desired and efforts and resources should be directed towards increasing the attendance of such students.

Gender distribution within CIoA FSP was not balanced (Table 6.2). The vast majority of students were female, while males consist of less then one third. Anecdotal evidence suggests that the same pattern is seen elsewhere (see Chaps. 3 and 4). These numbers also reflect similar patterns observed among all US students studying abroad. In its 2009 Open Doors Report, the Institute of International Education (IIE) reported gender distribution of 65% female and 35% male (from data collected between 1998 and 2008; see Table 24).

It is outside the scope of this chapter to provide extensive discussion for the reasons why significantly larger number of North American female students attends study aboard programs in general and archaeology field schools in particular. There are no studies of the issue and evidence is anecdotal. Random interviews with dozens of students to examine this pattern did not reveal any significant higher commitment to archaeology among female students. CIoA FSP students who self-identified as an archaeology major demonstrated a similar proportion to that within the larger attending archaeology field schools population There was no evidence of an overwhelmingly higher commitment to archaeology by female students.

Some archaeologists speculate that as the discipline matures, its historical gender imbalances are affected. As archaeology shifts from being a male-dominated discipline to being more gender balanced, it attracts ever-increasing number of females (for some discussion, see Claassen 1994; Hamilton et al. 2007; Reyman 1992). In addition, the intense competition over academic and professional positions in archaeology, coupled with declining pay, may be a factor in this gender shift. Regardless of the reasons, the high proportion of female students in archaeology field schools dictate changes in the way field schools are managed and run. Logistics must change to accommodate the gender imbalances. For example, accommodation is often separated by gender and when most attending students are female, securing equal-sized spaces – one for males, one for females – is no longer practicable. Instead, accommodations must take into account the gender imbalance and accommodate these patterns into the project logistical design.

Many directors anecdotally suggest that female students tend to be more patient, meticulous excavators. When the number of female excavators increases, so does the number of finds, especially small- and microsized finds. Projects must prepare for the increased data input and adjust excavation methodology to reflect the increased representation of small finds.

The decision to attend an archaeology field school usually comes late at a student career. Almost half the students who attended the CIoA FSP field schools each year were Seniors (Year 4), with Juniors (Year 3) as the next most frequent cohort (Fig. 6.3). Few Freshmen (Year 1) or Sophomores (Year 2) attend archaeology field schools. This pattern probably reflects the overall process of academic growth as Freshman and Sophomore students are still adjusting to live away from home, while Juniors and Seniors feel more independent and are better ready to explore the world beyond their home campus.

The academic standing patterns for participants on CIoA FSP field schools were dissimilar to those of other types of study abroad programs. The 2009 Open Doors Report (see Table 24) indicates that study abroad general population is usually 35% Juniors and only 20% Seniors. These differences suggest that archaeology field schools generally appeal to a more mature student population. Archaeology field schools usually take a much smaller number of students compared with the typical Travel/Study program and also are usually outside urban centers where fewer services are available and where long working hours of physical labor are required. Such conditions seem to appeal more to students later in their academic career, when they are ready to embrace the relatively more demanding conditions of archaeology field schools.

Annual enrollment for CIoA FSP commenced on November 15. Such early enrollment dates were required to allow students ample time to make plans, take

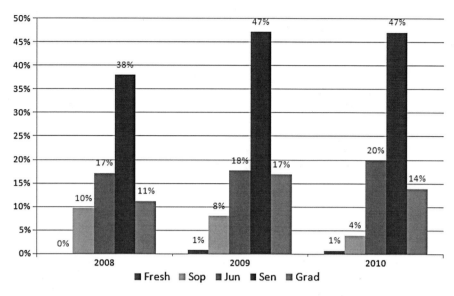

Fig. 6.3   Student standing among CIoA field schools program participants

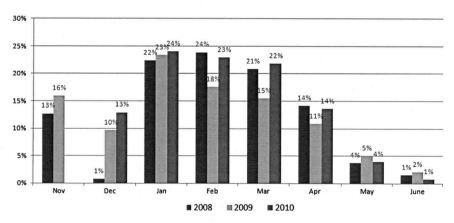

Fig. 6.4   Enrolment sequence to CIoA field schools program (enrolment for the 2010 season began on December 15 due to internal UCLA discussions concerning final program tuition)

classes, and apply to Financial Aid assistance in preparation for their participation. Students who wished to secure a place in a specific field school – due to popularity of location or desire to study a specific region, time period, or specialization – enrolled early and created a minor enrollment peak in November (Fig. 6.4). Enrollment subsided during December as students took their winter vacation and discussed with families their plans for the summer. Peak enrollment was between January and March, with a deadline for field school applications usually set by early

April, as were the application deadlines for Financial Aid, fellowships, and grants. Enrollment significantly declined in April and only few students enrolled through May and June. The late enrollees were usually students who could afford to field school costs without outside assistance.

A number of marketing venues and techniques were trialed by CIoA FSP. Program staff attended travel fairs in various universities, published advertisements in university newspapers, bought space on related websites, and used social media to advertise its field schools. Most of these efforts proved ineffective. Only three marketing tools emerged as viable paths to advertise to potential students. The CIoA FSP's own website and links from aggregated archaeology field schools websites (see below) were highly effective ways to recruit students. In addition, publicizing to colleagues in other universities also proved a successful method of advertising.

One of the most surprising results was the minor impact field school alumni had on enrollment. Given the very high satisfaction levels reported by students in their evaluations (see Table 6.1), it was anticipated that alumni will be enthusiastic ambassadors to the program. However, few students report that they heard about the program from a former participant. There are no satisfactory explanations to these results and it may be that archaeology field schools are viewed as challenging programs that even satisfied graduates cannot convince their peers to consider. The wide geographical catchment area of origin universities and single enrollment from the many participating schools – 111 universities contribute 171 students in 2010 – may support this assumption.

More in line with expectations was the role peer faculty played in enrollment. Students seeking quality summer programs ask for advice from their teachers, and if the teachers know about a program and respect the field school or the university running it, they are more inclined to recommend it to their students. UCLA enjoys a high ranking among US universities, and the CIoA is considered one of the leading global Institutions of Archaeology, and this reputation was certainly a factor influencing the ability of CIoA FSP to attract and recruit students.

Students also found CIoA FSP on the internet, though its own website did not tabulate the number of hits or the paths visitors took to find the site. It is clear, however, that students found the website either through research using web search engines or through reference sites that list archaeology field schools. Among the most popular are Shovelbums.com and the American Institute of Archaeology (AIA) sites. These sites consolidate listings of many archaeology field schools and the AIA site is particularly easy to navigate, explore, and compare programs.

Social media proved highly ineffective in marketing efforts. Despite having a CIoA FSP page on Facebook and MySpace, and having individual Facebook pages for many of its field schools, very few students found about the field schools through this route. Given the spectacular growth of the medium, it is surprising how ineffective it is in marketing field schools. It is possible that students are so saturated with "suggestions" from social media sites that they simply filter those out and ignore such recommendations. That former students are poor recruiters to the field schools further supports this hypothesis. Social media are built on trust and familiarity, and participants may ignore unknown or little known recommenders and view them as unworthy of their trust.

Mass email campaigns proved to be only of marginal effect. Although originating from trusty resources – in the case of CIoA FSP, the emails originated from the university's International Education Office itself – they were mostly ignored by students. It is possible that very few emails were seen as they were filtered through email provides to the "junk" boxes.

6.5 Conclusions

Archaeological research is about the study of past cultures that are significantly different from our own. It is a complex and sophisticated operation and frequently requires substantial planning and execution. Adding a field school element creates further complications that may affect the archaeological research itself (see Chaps. 2 and 7). Archaeology field schools require that time be allocated for teaching and for dealing with students who are usually young and for whom living in the field and in foreign locations is new. The inclusion of a field school slows down the research effort and necessitates other adjustments by project staff. However, the field school also brings a number of rewards, usually consisting of a significant funding stream and a capable and enthusiastic labor force (Walker and Saitta 2002:200), and helps to fulfill archaeology's central mission to educate the public about the past.

Each field school has its own set of interactions with and reactions to the local cultures among which the program is taking place. Authors in this volume discuss the benefits, tensions, and predicaments associated with this on-the-ground experience (see Chaps. 7 and 14). All CIoA FSP field schools dealt with these issues and each had its own methods to address them. There is no standard response as each situation is unique, influenced both by the community living around the project and the personalities of project director(s) and staff.

The field school can be viewed as a cultural phenomenon with shared characteristics that may be documented, studied, and analyzed. CIoA FSP was the first attempt to provide an alternative structure where the economy of scale allowed a significant shift in the way field schools were run, but also provided data to allow analysis of the phenomenon beyond the single field school. Because of the unique fee structure at the University of California, significant amounts of summer tuition were allocated to directly support field research. The Regents of the University of California wished to expedite graduation rates at its ten public universities and to open opportunities to the ever-increasing student population. They therefore created an attractive fee structure for summer sessions to encourage students to graduate early. Likewise, they encouraged departments to create classes for summer by allowing for a significant return from tuition to departments to spend on their own priorities. It is this system that has allowed CIoA FSP to direct most funds collected from tuition into direct and indirect support for its field schools. Aggregating many field schools allowed for increased efficiencies, could standardize minimum requirements, and created a brand name from which students could choose a program to their liking, trusting the "brand" to deliver quality experience.

The scale of CIoA FSP also allowed the accumulation of data and the analysis of trends. By examining the overall patterns from across the range of field schools, analysis of the CIoA FSP may allow the ways in which field schools are organized and managed to be adapted. It also invested in instruments that could be shared among the programs and negotiate terms to reduce costs of analysis. For example, CIoA FSP research projects could obtain AMS dates for a $210 a sample, a deep discount on normal market prices.

It would be desirable if the CIoA FSP model were applied elsewhere. While this may be problematic in the current economic climate, creating similar structures in other universities would only benefit archaeology as this structured access to field schools would attract even more students to the discipline and will push quality standards even higher.

The raw data presented here will also stimulate more research into the archaeology field school phenomenon, a purpose of this volume as a whole. Archaeologists need to think more about field schools as a primary teaching tool that deliver many benefits, even if most participants will not pursue a career in the profession (see Chaps. 2, 3 and 5). Field schools are playing an increasing role in building a future audience to the discipline. In times of rapid technological, economic, and political changes across the world, stewardship of the past may be problematic and the threat to archaeological remains in many contexts is increasing. Field schools may not be the only solution, but they can certainly play a significant role, especially if archaeologists will embrace the desire of students to participate not specifically in archaeological research of a particular site, but in a meaningful education experience that will allow both intellectual and personal growth.

6.6   Epilog

When this chapter was first written, the CIoA ASP was thriving and plans were laid to grow into 80 field schools annually. Strong support among UCLA academic faculty and high student satisfaction rates seemed to put the program as a shining example for the ability of a leading public university to fulfill its mission and vision. The program was financially self sufficient and was a leading example at UCLA for the entrepreneurial spirit advocated all across the campus. Alas, the program's rapid growth and sweeping success proved to be its Achilles heel.

By 2009, most of the field schools were directed by scholars, and attended by students, not organically from UCLA. Parts of the UCLA administration felt that the inclusion of these scholars and students under the UCLA liability umbrella exposes the university to undue risk. Students frequently become sick, and sometimes injured, in archaeology field schools and that concern cannot be simply dismissed. Litigation is always a risk – especially in California – and at the end of a long debate, UCLA made a decision that the liability exposure that the CIoA FSP brought was too great to tolerate. The decision was made to shut down the program, and the CIoA FSP was closed on December 2010.

Risk – and its management – must always be seriously considered when running any academic program. Of course, the only way to fully mitigate risk is to do nothing, and consideration for any program must balance exposure and mission. Most US universities are self-insured, resulting in highly avers tendencies towards risk. This is particularly noticeable in a time of significant financial difficulties, where universities are attempting to reduce costs and eliminate any potential for unforeseen payment due to ligation.

How can universities, then, pursue their mission? The dramatic shift in university focus to large financial awards is usually translated into investment – and tolerance of risk – in the so-called "Big Science." University administrations advocate projects and disciplines that promise huge financial return and are ready to tolerate risk there. Archaeology – even the CIoA FSP relative large scale – is a subject too small to attract support from most university administrations.

Although the CIoA FSP was a successful experiment, it was still too small to overcome internal institutional resistance. But its closure does not mean that the concept or the vision were invalid and should be allowed to peacefully perish. Instead, lessons learned at the CIoA FSP should be used to provide for the creation of a better, more stable organization that will be able to thrive despite the fear of risk.

If universities find archaeology field school programs too small to allow them to succeed, then the solution lays outside of their realm. In January 2011, I created the Institute for Field Research (IFR), an independent academic nonprofit organization dedicated to the development, running, and management of archaeology field schools around the world. Instead of being self-insured, IFR liability is managed through the purchase of specialized insurance policy that mitigates risk at reasonable cost. And because the IFR is run by archaeologists and for archaeologists, debates about risk and compliance with a huge range on internal university rules, regulations, and administrators are almost completely eliminated.

It is difficult to predict whether the IFR will succeed where the CIoA FSP failed. The new structure and experienced gained provide for a good foundation for success; there is ample reason to be optimistic.

References

American Council on Education. (2009). Letter to members of the US House of Representatives. Retrieved April 24, 2011, from http://www.nafsa.org/public_policy.sec/public_policy_document/study_abroad_1/foreign_relations_authorization/

Baxter, J. E. (2009). *Archaeological field schools: A guide for teaching in the field*. Walnut Creek: W. Left Coast Press.

Bender, B., Hamilton, S., Tilley, C., Anderson, E., Harrison, S., Herring, P., Waller, M., William, T., & Wilmore, M. (2007). *Stone worlds: Narrative and reflexivity in landscape archaeology*. Walnut Creek: Left Coast Press.

Buchli, V., & Lucas, G. (Eds.). (2001). *Archaeologies of the contemporary past.* New York: Routledge.

Claassen, C. (Ed.). (1994). *Women in archaeology.* Philadelphia: University of Pennsylvania Press.

David, N., & Kramer, C. (2006). *Ethnoarchaeology in action.* Cambridge: Cambridge University Press.

De León, J. (2010). *Undocumented migration project: Preliminary results of a study of modern undocumented migration through archaeology and ethnography.* Paper presented at the 75th anniversary meeting of the society for American archaeology, St. Louis, Missouri.

Fagan, B. (2006). So you want to be an archaeologist? *Archaeology Magazine, 59*(3), 59–64.

Gifford, A. C., & Morris, E. A. (1985). Digging for credit: Early archaeological field schools in the American Southwest. *American Antiquity, 50*(2), 395–411.

Gomarasca, M. (2009). *Basics of geomatics.* New York: Springer.

Hamilton, S., Whitehouse, R., & Wright, K. I. (Eds.). (2007). *Archaeology and women: Ancient and modern issues.* Walnut Creek: Left Coast Press.

Harrower, J. M. (2010). Geographic Information System (GIS) hydrological modeling in archaeology: An example from the origins of irrigation in Southwest Arabia (Yemen). *Journal of Archaeological Science, 37,* 1447–1452.

Haury, W. E. (1989). *Point of pines: A history of the University of Arizona Archaeological Field School.* Tucson, Arizona: University of Arizona Press.

Hodder, I. (2005). Reflexive methods. In D. G. H. Maschner & C. Chippindale (Eds.), *Handbook of archaeological methods* (Vol. 1, pp. 643–672). Walnut Creek: Altamira Press.

Joiner, C. (1992). The boys and girls of summer: The University of New Mexico Archaeological Field School in Chaco Canyon. *Journal of Anthropological Research, 48*(1), 49–66.

Kavanagh, F. B. (2003). *Geomatics.* Upper Saddle River: Prentice Hall.

Li, J., Zlatanova, S., & Fabbri, A. G. (Eds.). (2007). *Geomatics solutions for disaster management* (1st ed.). New York: Springer.

McManamon, P. F. (1991). The many publics for archaeology. *American Antiquity, 56*(1), 121–130.

NAFSA. (n.d.). Senator Paul Simon Study Abroad Foundation Act. Retrieved April 24, 2011, from http://www.nafsa.org/public_policy.sec/commission_on_the_abraham/

Perry, E. J. (2004). Authentic learning in field schools: Preparing future members of the archaeological community. *World Archaeology, 36*(2), 236–260.

Perry, E. J. (2006). From students to professionals: Archaeological field schools as authentic research communities. *The SAA Archaeological Record, 6*(1), 25–29.

Piscitelli, M., & Duwe, S. (2007). Choosing an archaeological field school. *The SAA Archaeological Record, 7*(1), 9–11.

Pyburn, K. A. (2003). What are we really teaching in archaeological field schools? In L. J. Zimmerman, D. Vitelli, & J. Hollowel-Zimmer (Eds.), *Ethical issues in archaeology* (pp. 213–223). Walnut Creek: Alta Mira Press.

Reyman, J. E. (Ed.). (1992). *Rediscovering our past: Essays on the history of American archaeology.* Brookfield: Avebury.

Staniforth, M. (2008). Strategies for teaching maritime archaeology in the twenty first century. *Journal of Maritime Archaeology, 3*(2), 93–102.

Tilley, C. (1994). *A phenomenology of landscape: Places, paths, and monuments.* Oxford: Berg.

Tilley, C., & Bennett, W. (2007). *The materiality of stone: Explorations in landscape phenomenology* (Vol. 1). New York: Berg.

VanderVeen, J., & Repczynski, J. (2010). The need for stewardship training in archaeological field schools. *The SAA Archaeological Record, 10*(1), 26–28.

Walker, M., & Saitta, D. J. (2002). Teaching the craft of archaeology: Theory, practice and the field school. *International Journal of Historical Archaeology, 6*(3), 199–207.

Wolf, R. P., & Ghilani, C. D. (2006). *Elementary surveying: An introduction to geomatics.* Upper Saddle River: Prentice Hall.

Part II
Teaching and Researching

Chapter 7
Two-Centre Field Schools: Combining Survey and Excavation in Ireland and Wales or the Isle of Man

Harold Mytum

7.1 Introduction

Field schools operate to complete a variety of pedagogic and research aims, as the diversity of approaches and solutions revealed in the volume testify. The two-centre field school model outlined here developed out of a long-running research excavation at Castell Henllys Iron Age fort and adjacent Roman-period native settlement which operated from its first summer season as a British training excavation in 1981. This was initially for University of Newcastle students, but once I moved to the University of York this institution became the main source of students. However, as the excavations became established and the camp site infrastructure was improved, students from other British Universities attended, as well as those still at school but considering archaeology, and also interested older participants. This arrangement was at the time a typical British equivalent of the North American field school, with its variable lengths of student attendance, less formalised teaching, and no assessment.

Research at Castell Henllys incorporated experimental archaeology from the beginning, as part of the site owner's public interpretation objectives which involved the on-site reconstruction of excavated structures (Mytum 1999a). For several years, Earthwatch volunteers took part in this activity alongside some of the students. The Earthwatch contribution also allowed the development of a parallel historical archaeology research programme on abandoned cottages and farms and on recording graveyard memorials belonging to the many different religious denominations present in the region. The project organisation therefore became structured in a way that allowed for a varied programme of activity at a number of simultaneous locations. This at times was logistically challenging, but did allow choice of student opportunities and a diversity of experiences. The Earthwatch programme ceased

H. Mytum (✉)
Centre for Manx Studies, School of Archaeology, Classics, and Egyptology,
University of Liverpool, Liverpool, UK
e-mail: h.mytum@liv.ac.uk

H. Mytum (ed.), *Global Perspectives on Archaeological Field Schools:*
Constructions of Knowledge and Experience, DOI 10.1007/978-1-4614-0433-0_7,
© Springer Science+Business Media, LLC 2012

after a number of years, but activity on the graveyards continued alongside the Castell Henllys excavations. A number of North American students had attended the training excavations, but at this stage there was no dedicated programme for credit. The British students attended as part of course requirements of their institutions that, at that time, were not measured in credit systems, though this has largely changed in recent years and has implications for British field training, partly discussed below (see also Chap. 3).

With a well-established fieldwork programme operating in Wales, a field school for North American and other overseas students for credit was designed that incorporated 2 weeks survey in Ireland, followed by 4 weeks excavation-dominated work in Wales. The programme timetable, its learning aims and objectives, forms of assessment, and credit value had to be approved through the University of York's administrative systems, though it was not able to issue official transcripts for this course as at this stage the University did not produce transcripts even for its own students. Instead, an official letter stating the credit value and grade was produced for each student. The field school was able to commence in the summer of 1995. With the transfer of the programme to the University of Liverpool when I moved to a new post, and the modification of the location to an Ireland and Isle of Man project focus, it has proved possible to construct a programme that can be provided with a transcript, though this can still prove problematic in transfer to students' home university's records, a problem that seems widespread with much debate concerning the transfer of credit values across national boundaries, even with established exchange programmes (e.g. Gold 2008; Kratz 2008).

The basic structure of the field school has remained constant through time. It consists of three sections, each of 2 weeks' duration, each with a focus on a particular stage of the archaeological process. The first 2 weeks, those in Ireland, concentrate on survey; the second element emphasises excavation; and the third has a focus on the collection, ordering and interpreting of archaeological data within a case study project that forms part of the fieldwork project of that season. The intention of the field school is to demonstrate the process of archaeology from reconnaissance, through excavation, to analysis and interpretation. To reinforce the stages of archaeological fieldwork, students produce a written assignment at the end of each 2-week segment of the field school. In each case, the intention is that students have to situate (and critically use examples from) their own field experience against the wider methodologies of that fieldwork stage, and the cultural context relevant to the period and place being studied.

Arriving at a field school where none of the staff or other participants are known, and in a foreign country, is a daunting experience. It is therefore important to allow everyone to settle in and relax, but before eating an early evening meal on the arrival day, we have a short meeting. At this introductory session, everyone (staff and students) first introduces themselves, and all the staff roles are explained. We then set out the house rules for rotas, sharing facilities, and the expected patterns of behaviour (including any necessary cultural issues that require addressing, such as the problematic nature of political and religious discussions in public in a cross-border region of Ireland). Some students have already been in Europe for a while, but others are still quite jet-lagged and are easily overwhelmed, so most of the staff then leave and I attempt to understand where everyone has "come from" intellectually, and what they hope to gain from the field school. Students at this stage are generally reticent on both their prior learning of methods, theory, or cultural archaeology, and vague about what they hope to achieve from the field school experience. Nevertheless, this does provide some pointers to how to best link their new experiences back to their past training. Little else is said at this stage to prevent overload, but we can consider how best to continue orientation and where any potential difficulties may lie. Anticipation of student stress points is vital.

Only a few students have taken part in archaeological fieldwork previously, and they all anticipate that what they will experience will be something very different from anything they have seen before. While many students admit to taking a basic field methods course, when theory is raised very varied experiences are offered. I have found that those interested in theory find its application within graveyard studies much easier than in most archaeology, as already data and contexts are readily available without the need of extrication from the earth and ordering and cataloguing (see also Chap. 8). For those with limited experience (and some basic theory texts are available in the reading for anyone wishing to start), some of the ideas can be illustrated through the material that they collect and can see across the graveyard. It is reactively easy to demonstrate contrasting interpretations of the same monument or monuments according to different theoretical positions, revealing to students how the same (or different) elements of the potential data are brought to play, emphasised, and used in various arguments. For some, this is exciting, for others worrying; they had hoped that fieldwork was unproblematic, straightforward, and led to empirical data that were incontrovertible. Facing up to such dilemmas early on, and with material that is relatively easily identified, is a real advantage in this first 2 weeks. Some students can carry this through to the later phases, but for others the methods of excavation and recording themselves dominate their attention, but at least all have had this aspect of archaeology revealed, and they can see that it does affect what is done, how, and what is made of the data collected through their efforts. This is reinforced in their last 2 weeks when they work on a project of their choice.

7.2 Part 1: Survey in Ireland

The Irish section with its focus on survey has two main components. The first is historic graveyard monument recording, and the second is surface survey. In some seasons it has also been possible to incorporate geophysical prospection and survey, but where this has not been possible this has been conducted alongside excavation in the remainder of the project. Occasionally, building recording has also been included in the survey element. Survey is chosen for Ireland for both research and teaching purposes, but also for logistical ones, as there are no recovered finds and samples that require storage or export permit, nor is the range of equipment too large for easy transport across the Irish Sea.

Fig. 7.1 Learning to set up the survey instrument for the graveyard survey

Surface survey is of historic graveyards, including the creation of plans of all physical features including walls, buildings, monuments and paths, the vegetation, and contours revealing the topography. This has been with an EDM or total station (Fig. 7.1), though the students have generally written down the coordinates and plotted them out rather than using software as this reveals more of the process and reinforces the spatial relationships that are being recorded by the mapping. Having a mixture of points on "hard" features such as the corners of tombs, and readings taken to define "soft" features such as extent of vegetation or to reveal slopes, makes the graveyard an ideal location for survey training. It also often requires relocating the survey station to new positions as only some of the many features are visible from any one point, giving students experience in setting up and taking down the instruments to create a complete plot. Surveying sites that are also being recorded in other ways also encourages the students to think about the value of each type of evidence being collected, and how the combination of sources allows issues to be addressed that could not be considered with only one category of data. The surface surveys are similar to those carried out at many archaeological sites, though they tend to contain more features and at a greater density than is normally the case for surface survey.

The graveyard recording process follows a methodology (Mytum 2000) which creates data of a standardised format ready for analysis. Indeed, this recording structure evolved in stages over the field school seasons, with feedback coming from students recording and using the data, as well as from subsequent analysis to produce academic output. The details of the recording vary slightly from season to season depending on the research questions though, as with British excavated context recording, there is a core set of data that is seen as essential for others' use that is always collected. The record forms include information that in some senses is comparable to many artefact records, with measurements, codes for condition, materials, forms, surface treatment and decoration. There are additional fields, however, including orientation, nature of the inscriptions, and language, as well as a transcription of the text. All memorials are also sketched, as this forces the students to really notice the form and decoration. Some drawings are truly excellent, with some more artistic and others more technical in style. Other drawings can be relatively crude, but as long as the essence of the monument has been captured, this is still satisfactory, as photographs are also taken of every monument, and for many analytical purposes, these are the primary visual record. Thus, there is no preferred drawing style for the graveyard forms, unlike for the excavation drawn record discussed below, where clear standard conventions apply.

In the early years, the photographs had to be printed out and affixed to the forms, but now digital images allow a different method of recording, with much more rapid appraisal of the quality of student photographs, and the greater ease of taking multiple images of the same monument. It is also easier for the students to see if their compositions are effective, as the images can be reviewed immediately, though only when downloaded can photograph focus and the legibility of the texts be better evaluated. It is also possible to have all students take a significant number of photographs; the less effective duplicates can eventually be discarded from the final project archive.

While historic graveyard monuments might seem an eclectic and marginal archaeological data category, many of the recording processes are similar to those for other categories of data. Moreover, the monuments themselves are easy for students to comprehend, they rapidly become accustomed to the codes and ways of filling in the forms and, after a few days, many become expert at reading even worn inscriptions and symbols. This gives confidence and encouragement to the students, something that often takes longer if the first exposure to fieldwork is excavation to reveal the subtle traces of differential soil colours and textures. The concepts of codifying and measuring data, transferring from a physical, three-dimensional object to a set of data on a form, can all be effectively taught using graveyard monuments (Fig. 7.2).

Most students do not sign up to the field school for the graveyard component, but recognise the importance of survey. Once in the field, however, many become enthused by this aspect, and some have continued this interest back home, carrying out independent research projects for credit at their home undergraduate universities, or developing such a research theme for subsequent Masters dissertations. The relatively rapid production of research papers based on field school data allows students to see how their work will be used (e.g. Mytum 2002, 2004a, b, 2006) and also the issues in those papers can be identified by students as they become used to the recording process. A very small minority have considered the subject either morbid or not "proper" archaeology, though such an attitude was often already fostered by

Fig. 7.2 Graveyard recording starts requiring much staff support, but students gain confidence and experience and are later able to work independently for some of the recording tasks, and in teams for others

a rather "macho" view of archaeology as just digging, and all other aspects were treated with disdain. Such an attitude to archaeology seems extremely outdated, given the amount of landscape and buildings archaeology, as well as heritage management, that now occupies a significant part of the profession and which involves no excavation at all.

Over the 2 weeks, the students enter up the data into spreadsheets on laptops back at base and order and name the digital images to create a coherent archive. This often requires subsequent checking and sometimes further visits to the graveyard for missing data or to obtain improved photographs, but the vast bulk of the data entry is achieved during the field season. Some students really enjoy the data entry, while others find it tedious. However, all are impressed by the speed at which that some basic descriptive results from a graveyard can be produced, either at the end of the survey fortnight, or a short time into the excavation phase once the data have received a preliminary check by staff.

General and region-specific reading on Irish history, historic archaeology and graveyard studies is provided at the project base, and students are expected to engage with this material to contextualise their field experience and consolidate the interpretations offered by staff during the fieldwork. In some seasons, particular research themes are being highlighted in the fieldwork so that certain studies can be completed, such as ethnic and religious differences (Mytum 2009) or spatial patterning graveyards (Mytum and Evans 2002), while in other years samples are being accumulated to provide longer term research goals. Most students have received no training, even in class lectures, in the field techniques applied in this survey section of the programme, but the integrated format and the easily comprehended research aims and relevance of the collected data make it relatively easy for students to grasp the issues, and many produce excellent written assignments. This stage gives students the confidence to confront the physically more demanding and less immediately comprehensible evidence that comes from excavation.

Graveyard recording is non-destructive and extensive, with a field school crew potentially recording at more than one graveyard site in a season. As it is easy to return to complete a survey the following season, there is less pressure on completion of projects than in most fieldwork. Nevertheless, not all research objectives are long-term; certain data sets may be desired for a particular research paper to be delivered during the following year, and that can create tensions between keeping students busy, motivated, and developing their skills, and completing the recording of a particular site or category of data. One particular difficulty that arises is the relatively slow pace of mapping the whole graveyard compared with filling in forms for every monument. Photography normally keeps pace, but sometimes this can also be slow, or only possible for monuments in some parts of the site at certain times of the day or when the weather is suitable for reasonable quality images. Irish weather can impede progress, though the use of golfing umbrellas allows most forms of recording to continue in rain providing there is not a strong wind, but again this can skew an even progress on all fronts. As a result, a site can be finished apart from the measured plan, and so the logistics of transport have to be flexible enough to allow the team to split when necessary. Fortunately, within the range of research questions asked during the field season, not all require measured plans of every site, so it is possible for the priority sites to be mapped, but data with only a sketch plan obtained from other locations. Over the years, it has also become apparent that the data checking process requires experienced staff to work independently on checking data and correcting errors and filling in omissions, undisturbed by student questions. All field school recording requires a robust quality assurance mechanism, whatever the form of the fieldwork.

The survey phase of the project is based in rented accommodation, where the students share rooms and there is a project office where data are processed and the small library is housed. Various forms of shared cooking have operated over the years, but the standard of comfort is relatively high, and a short walk brings students and staff to a pub, and to shops. This phase of the field school consists largely of field school students and staff, usually with just a handful of students from British or Irish universities. This allows the students to get to know each other and the staff, and settle into routines and expectations. Cultural differences begin to become apparent, but the North American students form about half the number of people in the team, so they do not feel overwhelmed. For example, students make sandwiches on a rota basis for the team lunches, but the field school students' discovery of what constitutes a British or Irish-style sandwich always creates amusement on all sides, and the American, Canadian, British and Irish students all learn how they have variously overlapping vocabularies for many everyday items and behaviors.

The survey phase in Ireland works well for 2 weeks, but the transfer to a new location and different living conditions, as well as to new archaeological challenges, means that any simmering inter-personal tensions, or frustrations with any aspect of the work, are dissipated. The 2-week first stage is also very clearly finite, which encourages students to perceive time constraints in fieldwork and that delays caused by errors, bad weather, or equipment failure all have obvious implications. This length of time is sufficient for almost all students to become proficient and reliable data collectors, so they can see how they have progressed from their first day. This gives them confidence that, however difficult and confusing excavation may seem at first, they will be able to master those skills also in the time assigned for this.

7.3   Part 2: The Excavation

The students physically move location for the excavation, crossing the Irish Sea and arriving at a camp site where their individual tents have been set up. They no longer share personal space and have their own tent, but that space is of a form that some have never previously experienced. Moreover, most have not camped in the context of a mercurial climate, though their prior Irish experience in the previous 2 weeks has been a useful reminder that, unlike most North American summer weather conditions, being able to predict the weather for 2 hours is a luxury which should not be assumed. Another major change in the dynamics of the field school is that the students are joined by more British and European students, creating a different web of social interaction. This has a beneficial effect, as friendships already created are maintained, but less significant associations can be replaced by others formed within the wider excavation community.

The basic training of excavation and recording techniques is similar for all students, though special attention is given to providing the field school members with additional teaching. This largely concentrates on explaining the variety of methods used, the differences between various traditions within North America and how excavation is conducted in the UK (Barker 1977; Carver 2009; Drewett 1999). Some standard aspects of North American excavation, such as screening or the use of test pits, are included even when the research design typical of a British research project might not demand it to the same degree (Fig. 7.3), to provide appropriate experience (see also Chap. 12). While some of these differences might be confusing at first, the discussion and reading of British text books consolidates these techniques and invites consideration of what lies behind these different styles of excavation and recording. Instead of just learning methods by rote, students understand the relationship between research questions asked, the types of data deemed to provide

Fig. 7.3 Test pits and screening can take place on the excavation, even if not always required within a traditional British research design which often required open-area excavation

answers, and the methods by which those data sets can be acquired and recorded. The combinations of students working together are altered as they are gradually assigned new tasks, creating a fluid pattern of interactions that means that those few students that irritate others are shared between many, and those with close personal relationships can be kept separate while working on-site.

For many years, the scale of the Castell Henllys Field School allowed choices about the type of site on which to excavate, with later prehistoric (Mytum 1999b; Murphy and Mytum 2007) native Roman (Murphy and Mytum 2005) and historic (Mytum 2010) options available. Even when all these periods are not represented, there can still be choices between working in trenches or test pits or open-area excavations, though often students wished to experience a variety. The challenge in advising and managing the students was to ensure that they received the full range of training and also concentrated long enough on any one field task to become competent in it. Some students so enjoy some aspects they are unwilling to be moved on to another task that is to be learnt, while a smaller number have insufficient patience to stay with one activity long enough to fully appreciate its complexity and variability. While many field schools allow students to work on one test pit from beginning to end, which creates a coherent narrative, it also limits variety of experience. British archaeology tends to work on large areas and concentrates on features, so a variety such as post molds, defensive ditches, walled rooms and drains could provide a variety of different forms of feature to work on and consider. Recording methods include filling of context forms, planning, and profile drawing, and these can be interspersed with artefact processing and wet screening and flotation for environmental samples.

Field school students are always mixed on-site with the British students and other attendees, to create an integrated workforce where cliques are avoided and cultural interaction is maximised. Moreover, British students are mainly single subject archaeology students and have often undertaken far more courses in the subject than North American students, though they do not have the benefit of a broader curriculum. This creates much interest in each other's university systems, contact levels with staff, methods of assessment, and career ambitions. It also allows the field school students to find out about local cultural traditions and norms; they learn a great deal just by working together, greatly augmenting what is taught in explicit training by staff.

The field school students receive some additional training not experienced by the British and European students, as they have to be able to write a report on their excavation work at the end of the 2 weeks. Some find it difficult to read methodological and appropriate cultural background material with the distraction of the other students, but having their private space of their own tents, and for most years a dedicated communal study area for field school students, helps. The return of their provisionally graded first assignment with detailed written feedback helps to concentrate their minds on the next assignment and, most importantly, review what they have already learnt on the excavation. Students have time set aside to complete their assignments which combine evidence of their excavation and recording experiences in the context of the project research aims set against the wider cultural context of the site they have chosen. Reading is not the only source of information, as after-work trips to other sites in the region provide complementary evidence and a greater understanding of the cultural sequence, topography, and other archaeological fieldwork in the region. Longer trips are also arranged for the day off that combine opportunities for shopping and visiting a beach as well as seeing other archaeological and historic sites. Embedding the student experience within the wider regional and cultural context is extremely important; before they arrive, most students are at best only dimly aware of the Welsh or Manx languages, local architectural and agricultural traditions, or the national identities of these countries.

The field school creates an identity that can be reinforced by the production of T-shirts and social events that bind the whole group. Although only few field school students return, many British and other European students have come back for further seasons, to expand and develop their skills, meet old friends, and see the development of the project. The York undergraduates had the opportunity of taking one assessed unit where they took more responsibility in the field season and wrote a reflexive project report on this. Through this method and the selection of capable returning students from elsewhere, it has been possible to create a structured series of levels of experience that provides the core staff with assistant supervisors. By this means, a significant number of the Castell Henllys students have been able to progress in academic or CRM archaeology. The consideration of returning student skills and personality is an important part of the team-building on the field school. While occasionally a student has not matched up to expectations, in most cases the opposite is true, with a real step-change in confidence, motivation and skill over the years. I have found that acting as director for a dispersed and varied set of field-work activities, it is vital to create a hierarchy of responsibility, lightly but clearly in operation. Just as others have found gender imbalance in field schools (see Chaps. 3 and 6), with a significant majority of female participants, there has also been a tendency for women to dominate the finds processing, with only one male student in over 25 years selecting this as a supervisory role at Castell Henllys. In contrast, other aspects of the excavation (and indeed graveyard recording) has been much more evenly represented over time. Everyone needs to feel supported and appreciated in their role, with all receiving attention and recognition. It is thus important to ensure that those dealing with finds or environmental samples feel as much part of the project as the excavators on-site, and that those areas of the excavation producing less obviously exciting results realise that these all contribute to the overall picture.

7.4   Part 3: The Project

The final stage of the field school experience is the choosing of a small project and carrying it through to the production of a report. While students may select a graveyard, mapping or surface survey project, most desire to develop their excavation experience, and this is certainly easier to manage and support, as it forms part of the ongoing excavation. A particular site survey may be drawn up and interpreted spatially, and graveyard monuments might be examined for changes through time or class or gender differences. On the excavation, students most often select a particular part of the site that they have already found interesting and work on a particular complex feature or area. Using already extant records, they continue to excavate and record to gather more data and discuss with the area supervisor and myself what questions require answering from the plan and sequence that is being recovered. It is important to define a project that is sufficiently substantial that there is a question that can be asked but one that is not too complex or where the excavation will not be completed in time for the analysis and report to be written by the end of the 2 weeks. By the end of the fifth week, students receive back their excavation assignment with written feedback and a preliminary grade.

Fig. 7.4 Students work together on some aspects of their project, such as assembling an overall site map, but then each concentrate on their own analysis

The field school students usually stop fieldwork a couple of days before the end of the excavation in order to complete their reports, leaving the other students to continue on-site. The field school participants may also have already spent some time carrying out background reading, a preliminary review of the records, and closer definition of their project brief before this point. While this reduces the amount of field archaeology undertaken by these students in the last week, it greatly consolidates their understanding. Suddenly, the reasons for writing coordinates or stratigraphic relationships become important. They require that information for their own report, and no longer does it seem arcane detail that is written down to conform to a taught protocol. The engagement with a project encourages students to think more deeply as they excavate, and so builds upon rather than merely repeats the experiences of the previous 2 weeks of excavation; they no longer move about the site but become wedded to their own part of it and become engrossed in its problems. As the season draws to a close, they understand the pressures of time, the ways in which archaeology works through a series of stages that cannot be bypassed, and that all decisions can lead to the collection of more or less data at varying degrees of resolution (Fig. 7.4).

The project reports vary in word length as some rely on tabulated or graphical data more than others, but all set up a problem and address it with reference to primary records from the project, some of which they will have created themselves. The project is situated within a methodological framework which can be compared with others, and a particular problem is addressed through the primary record but related to what is already known from other excavated and unexcavated sites to provide some form of interpretation. While most projects are of necessity partial and students often make some errors in their assumptions because of their limited experience of both field archaeology and the culture that they are reporting on, it is pleasantly surprising how far many do get in making sense of the records and creating coherent argument and interpretation. Most students feel that the project forms an appropriate finale for their field school experience. They are able to enjoy the end of season festivities and head home feeling how far they have come in understanding, knowledge, and experience over their 6 weeks.

The project reports are graded and returned with feedback, so that students can see which aspects were most successful and know where they made mistakes and what alternative interpretations could apply. By this stage, the field school students have received all their preliminary grades, but in the UK system assessed work may be moderated by another member of staff, and this internal review now takes place in Liverpool before the university can begin to process these to create transcripts with the final grade. Our field school is more concerned with personal development and increasing understanding than it is with grades, though high marks come from deeper learning, so they are not unrelated. A few students may be obsessed with grades, and they need to be encouraged to participate and learn first, to then concentrate on the actual assignments when they draw near. Field schools are fully participatory and are quite unlike classroom learning; an eye only on written assignments does not maximise the experience and often is counter-productive. The field school assessment does not include actual competence on-site as the UK system requires the option of moderation and potentially scrutiny by an external examiner from another university. Moreover, grading of performance can be viewed by some students as personalised and open to favouritism, whereas the written assignments with detailed feedback, available should there be any challenge to the grading, indicate strengths and weaknesses in the students' submissions, from which they can learn for their next task. It is through personal supporting letters and references that practical competence and other personal characteristics can be mentioned, and as a field school director, I am frequently asked to support applications to graduate school.

7.5   The Balance of Teaching, Learning and Experiencing Across Two Locations

The different locations offer varied experiences which contribute to learning in both formal and informal ways. The shared experience of working together bonds students and staff, but the variations in the other student and volunteer participation over the 6 weeks creates variety and prevents an introversion that can develop within a relatively small group for many weeks on end. Some students have some ancestry in one or more of the countries visited, and this gives a personal relevance to the work there, though this is perhaps more important in the other experiences gained outside the formal working and learning environment. Visits to pubs and hearing local music, trips to museums and other cultural attractions, and the experience of the landscape, land use and local people create an impact which is more than is achieved by being a normal tourist.

The field school is in a location for a purpose, which has value and direct relevance to that local community. This gives confidence to the participants and also provides a point of engagement that the typical visitor does not have. Being in one region, out in the landscape day after day and carrying out what is visibly a form of work creates a bond with local people that tourists do not share. Visitors are seen as people to be welcomed but largely for financial gain through the provision of goods and services, but the field school members are active in the region, discovering things about its past. Even if archaeology is viewed as eccentric by some, it is largely seen as positive, creating a wider understanding of place and time. By carrying out work in two locations, different attitudes to the heritage can be revealed to the students, not through formal classes, but through their interaction with local people. They can discover how contingent is the interest, prior knowledge, and cultural value that is given to heritage in different places and among different groups.

Teaching in different locations has some logistical challenges, creating internal breaks that increase the number of deadlines, though this encourages self-control and prevents slippage from one aspect of the research or teaching to another. Moreover, the split of survey and excavation by moving location, and from mortuary context to settlement, creates a shift in student focus that introduces a new momentum. The few who have been less interested in graveyard survey are ready for the next stage, and the great majority that enjoy it have the confidence for a new challenge, and move on before any boredom sets in. Although the excavation lasts 4 weeks, it is also split into two stages, and having had the first discrete 2-week session, this is easily appreciated by the students. When working for 6 full weeks at Castell Henllys, it was easy for the team to settle into a quiet middle phase after initial enthusiasm and before the final wish to complete certain targets. Central season "drift" was always a challenge; this does not occur with the 2 weeks plus four, the latter perceived by all as multiple blocks of 2 weeks.

The numerous local archaeological sites available to visit means that less time is spent in formal classroom-style lectures than in some field schools because the regions offer more potential that should be offered to students from other countries. Students learn a great deal from on-site talks and discovering features of castles, forts, churches and prehistoric monuments themselves within a supportive learning environment of background reading and some expert guiding. Moreover, between the sites, and during the travel from one base to another, students see and have pointed out local architecture (historic and contemporary), land use, and animals. Some students have never seen the sea, a sheep, or a donkey. Students also learn life skills, sometimes linked to a different cultural context, but others are basic ones such as washing up dishes by hand or cooking other than with a microwave.

The learning on the field school relates both to archaeological methods and their relationship to theory and research designs and to the particular cultural histories being investigated, but much of the learning is about the worlds in which the student operates during the field school, and their discoveries about themselves and their relationships with others in various conditions. By having two complementary but significant different contexts within which the field school operates, this provides another level of comparison, and variations in the challenges for the students which they relish rather than fear. With a core staff that are present throughout and provide support, students gain experiences outside their own normal worlds, which is why they have signed up for the field school in the first place. It is possible to see students gain in confidence, but also appreciate the differences in political and cultural world views that can be held across Anglophone countries; they return noticing how Irish and British media report more on world events, and with a greater diversity of views than they are used to. They discover that North America is not the centre of everyone's world, and that simplistic nationalistic stereotypes do not fit real people. Just as field school staff realise that North American students are very variable in background, experience, opinions, and aspirations, so the students find out the great differences in educational systems, cultural heritage, and lifestyles, even in countries that many of them expect to be similar.

7.6 Conclusions

While many of the objectives set for the Castell Henllys and now the Ireland and Isle of Man field schools could be achieved in a single location, there is much to be gained by a two-centre model. Although the logistics can be challenging, and the costs higher because of the travel within the field school timetable, the advantages from a pedagogic and research viewpoint are considerable. The framework allows formal comparison of archaeological historiography and approaches in the different countries (and also with North America) and the experience of archaeology in different regions. It structures the learning into discrete sections that can be self-contained to the benefit of student and staff. The most important advantages, however, lie in the opportunities for greater independent, informal, learning about the past and the present, and about the students understanding of themselves and their peers as people. Most students leave recognisably transformed by their experience, though many of the effects are often only apparent to the students after their return and the ways in which they then relate to new courses in their subsequent career. Written student feedback at the end of the field school tends to emphasise learning through aspects of the formal training and the feedback on their assignments, which is in far greater detail than many have previously been given. Subsequent email comments or discussions when meeting up at conferences later in their careers suggest that students gradually become aware of some of the intangible learning outcomes, such as their increased awareness of cultural assumptions and agendas, and that there are many more ways of doing and thinking archaeology than they had first thought.

In the longer term, these may be perceived as the most significant outcomes of their experience. While the provision of technical competence is a given for field schools, it is what else they offer that really distinguishes them from the formal teaching that can be accomplished on campus, whether in the classroom, the laboratory, or even outside.

References

Barker, P. A. (1977). *Techniques of archaeological excavation*. New York: Universe Books.

Carver, M. (2009). *Archaeological investigation*. London: Routledge.

Drewett, P. L. (1999). *Field archaeology: An introduction*. London: UCL Press.

Gold, A. (2008). *Grade conversion and credit transfer for exchange students*. Paper delivered at the BUTEX international conference 2008. Retrieved February 28, 2011, from http://www. butex.ac.uk/?q=node/28

Kratz, A. (2008). *Credit or grade transfer. A case for grade transfer*. Paper delivered at the BUTEX international conference 2008. Retrieved February 28, 2011, from http://www.butex.ac. uk/?q=node/28

Murphy, K., & Mytum, H. (2005). Excavations at Troedyrhiw defended enclosure. *Archaeology in Wales, 45*, 92–94.

Murphy, K., & Mytum, H. (2007). Excavations at Berry Hill inland promontory fort, near Newport, Pembrokeshire, 2007: Interim report. *Archaeology in Wales, 47*, 82–88.

Mytum, H. (1999a). Pembrokeshire's pasts. Natives, invaders and Welsh archaeology: The Castell Henllys experience. In P. G. Stone & P. Planel (Eds.), *The constructed past. Experimental archaeology, education and the public* (pp. 181–193). London: Routledge.

Mytum, H. (1999b). Castell Henllys. *Current Archaeology, 161*, 164–171.

Mytum, H. (2000). *Recording and analysing graveyards*. Council for British archaeology handbook 15. New York: Council for British Archaeology.

Mytum, H. (2002). A comparison of nineteenth- and twentieth-century Anglican and nonconformist memorials in North Pembrokeshire. *The Archaeological Journal, 159*, 194–241.

Mytum, H. (2004a). Artefact biography as an approach to material culture: Irish gravestones as a material form of genealogy. *Journal of Irish Archaeology, 12*(13), 111–127.

Mytum, H. (2004b). A long and complex plot: Patterns of family burial in Irish graveyards from the 18th century. *Church Archaeology, 5*(6), 31–41.

Mytum, H. (2006). Popular attitudes to memory, the body, and social identity: The rise of external commemoration in Britain, Ireland, and New England. *Post-Medieval Archaeology, 40*(1), 96–110.

Mytum, H. (2009). Mortality symbols in action: Protestant and catholic early 18th-century West Ulster. *Historical Archaeology, 42*(1), 160–182.

Mytum, H. (2010). Biographies of projects, people and places: Archaeologists and William and Martha Harries at Henllys Farm, Pembrokeshire. *Postmedieval Archaeology, 44*(2), 294–319.

Mytum, H., & Evans, R. (2002). The evolution of an Irish graveyard during the 18th century: The example of Killeevan, Co. Monaghan. *Journal of Irish Archaeology, 11*, 131–146.

Chapter 8
Constructing New Knowledge in Industrial Archaeology

Timothy James Scarlett and Sam R. Sweitz

8.1 Introduction

Industrial archaeology (IA) is the study of the physical remains – the artifacts, systems, sites, and landscapes – of industrial society, including their cultural, ecological, and historical contexts. Practitioners of IA not only study these remains, but are also often involved in their practical preservation, management, and/or interpretation.[1] Over the last 50 or 60 years, IA has matured from its early beginnings in the United Kingdom into a worldwide interdisciplinary community of people drawn together by collective desire to understand the industrial world.[2]

---

[1] Seely and Martin (2006) have written a short history of the IA program at Michigan Tech that included the philosophical justification for our design of the Industrial Heritage and Archaeology Ph.D. Analysis and discussion of the Michigan Tech's M.S. degree, including comparisons to other programs in heritage or industrial history, were published by Crandall et al. (2003), Weisberger (2003), and Martin (1998, 2001).

[2] Industrial Archaeology began in England as a combination of scholarship and activism aimed at preserving or recording the earliest remains of the industrial revolution, and spread through the United Kingdom (Buchanan 2000; Palmer 2010; Palmer and Neaverson 1998:8–15) then quickly through the United States, Western Europe, Canada, Australia, and Japan. IA developed differently in various countries, but has generally been inclusive of avocational involvement through local societies and organizations. Martin (2009) recently overviewed the development and internationalization of IA, connecting it to many of the themes in this chapter, and situated the West Point Foundry project among them. Many IA practitioners have also published for audiences of enthusiasts along with their colleagues. Whenever and wherever IA found an academic home, it was often in adult education programs in a particular national college and university system (Martin 2009:286) or at newly formed open-air or eco-museums (Storm 2008:29–46). These trends led to periodic debates over how IA is to be defined, for example, whether it should be more or less tied with resource management and the heritage industry (Alfrey and Putnam 1992; Palmer 2000). Martin's (2009:286–289) overview included a review of IA's development in the United States and further

T.J. Scarlett (✉) • S.R. Sweitz
Department of Social Sciences, Michigan Technological University,
1400 Townsend Drive, Houghton, MI 49931, USA
e-mail: scarlett@mtu.edu; srsweitz@mtu.edu

H. Mytum (ed.), *Global Perspectives on Archaeological Field Schools:*
Constructions of Knowledge and Experience, DOI 10.1007/978-1-4614-0433-0_8,
© Springer Science+Business Media, LLC 2012

We have been collaboratively teaching IA field schools at Michigan Technological University for 10 years. Tim Scarlett joined the Michigan Tech faculty in 2001, and for 10 years before that, he had taught field archaeology at industrial sites. Sam Sweitz began teaching in the IA program in 2005, and also has a similarly long interest in industrial heritage. We are both anthropologists trained in the American style of a four-field approach, which ties archaeology and ethnography with linguistic and biological anthropology. We are also Americanist scholars in that our research has concerned the industrial history and cultures of North and Central America, and the Caribbean.

Michigan Tech's Department of Social Sciences has offered an annual field school in IA for more than three decades. We offer the field school in conjunction with our graduate degree programs[3] and the majority of our field school students during the last 10 years have been enrolled at Tech pursuing degrees in either Industrial Archaeology (M.S.) or Industrial Heritage and Archaeology (Ph.D.). The graduate programs in the Department of Social Sciences are unique in North America, and given the interdisciplinary design of our program, our students undertake courses of study unlike any others in the world. Our program blends scholars

---

overviews or case studies can be consulted for Sweden and Scandinavia (Nisser 1983), Europe (Palmer and Neaverson 1998:8–15), Australia (Casella 2006), Japan (Komatsu 1980), as well the spread into Mexico and Latin America in the 1980s and 1990s (Oviedo 2005, and the rest of *Patrimoine de l'industrie/Industrial Patrimony* 13, Part I: 7–66) and Spain (Cerdà 2008). Published field guides and inventories of industrial heritage are very numerous. These national and regional movements were united in the first International Congress for the Conservation of Industrial Monuments in Ironbridge, England, in 1973. In 1983, delegates from many nations meeting at the third international congress established The International Committee for the Conservation of the Industrial Heritage (TICCIH). There remains a strong distinction between the Anglophone traditions of industrial archaeology in England, the United States, and Australia, and the contrasting idea of *Industriekultur* in continental countries like Germany (Ebert and Bednorz 1996) and Sweden (Storm 2008), as well as the traditions of Iberioamérica (Areces and Tartarini 2008). The nascent involvement in TICCIH by representatives from India and China (Dong 2008; Joshi 2008) will add more distinctive voices to the community. Industrial Heritage is flourishing around the world, a fact made clear by the many excellent publications like the journals *Industrial Archaeology Review*, *IA: The Journal of Industrial Archaeology*, *Patrimoine de l'industrie/Industrial Patrimony*; bulletins of professional and avocational societies, such as the *TICCIH Bulletin*, and the creation of numerous industrial museums, monuments, landscapes, festivals, and heritage areas now busily being organized into ever larger networks of industrial heritage like the European Route of Industrial Heritage (http://www.erih.net).

[3] Michigan Tech began accepting graduate students to study for a Master's of Science degree in Industrial Archaeology (M.S.) beginning in 1991 as well as a Doctor of Philosophy in Industrial Heritage and Archaeology (Ph.D.) in 2005. While the Department of Social Sciences has always had a small number of undergraduate students studying for degrees in history, social sciences, or the teaching credential associated with those degrees, the department only recently created an undergraduate major in Anthropology in 2009. The addition of this degree seems to have also caused an increase in the number of undergraduate field school enrollees, but more time is needed to know if the intellectual balance of our field schools will change. Information on all the graduate degree programs as well as details on planned field schools can be found at http://www.industrialarchaeology.net.

Fig. 8.1 Industrial archaeologists must be able to collaborate with members of descent communities. Graduate student Carmelo Dávila interviewed José Ramón Rivera about the community and his work as a sugar mill employee in Aguirre, Puerto Rico, in 2007 (photo by Sam Sweitz)

and perspectives from IA, historical archaeology, history of technology, ethnography and social history of industrial communities, material culture and architectural history, heritage management and documentation, all of which are unified through a heavy focus on field training with the material remains of industry (Fig. 8.1).

We expect that all the graduate students in our program should learn more than theoretical justifications for IA, but also master basic IA fieldwork skills, including recording historic structures and conducting archaeological excavation. Our students must combine the skills of documentary photography and measured-drawings[4] with common techniques of archaeological excavation and the scientific tools that have come to be expected of modern field archaeologists: digital total station survey, Global Position Systems, Geographic Information Systems, AutoCAD, and experience with some type of remote sensing or imaging technology.

---

[4] We teach photography and drawing as part of our regular curriculum using the Secretary of the Interior's Standards and Guidelines for Architectural and Engineering Documentation. In the United States, National Park Service's Heritage Documentation Program administers the Historic American Engineering Record, Historic American Building Survey, and Historic American Landscape Survey (collectively known by the acronym HABS/HAER/HALS). These policies are available at: http://www.nps.gov/history/hdp/standards/index.htm.

At Michigan Tech, we introduce our students to IA as a global field. Practitioners in this global IA community have widely varied relationships to the definitions of archaeology created through academic or government bureaucracies, as one would expect. As the junior scholars at our institution, we are pushing IA to grow beyond the traditional "core" of the field, exploring alternative regions, industries, forms of production, and perspectives. Yet we do this while preserving the traditional strengths of the field, with its focus upon the evolving technologies and social networks of production.

Hardesty (2000) wrote of the "voices" of IA. In his essay, he discussed the overlapping and distinct communities that participate in fieldwork. Our thoughts on the experiences of IA field schools undoubtedly reflect our backgrounds as anthropologists (and one of us studied under Don), but we think that our experiences meaningfully connect to larger issues and concerns within many academic training programs. Building on the idea of voices, we introduce each section of this essay with a quotation. These words were spoken by our students, our colleagues, or one of us during a field school or class activity. Occasionally, we have been forced to paraphrase or soften a student's word choice, but have retained the spirit of their thoughts.

8.2   Wait, Hold On: You Get to Do Mechanical Drafting *and* Archaeology?

This fall term, one of us ran into a young woman in our campus library coffee shop. She had enjoyed Tim's general education lecture course for first year students and had stopped him to ask what he was teaching this term. He told her that he had just started teaching our department's IA course and that over the weekend, the class had hiked out into the mountains to measure and draw some midnineteenth century stamp mill machinery that still lay *in situ* at a mine site here in Michigan's Upper Peninsula.

In her excitement, Tatiana had interrupted Tim midsentence to ask her question about the intersection of mechanical drafting, engineering, and archaeology. The incredulous look on her face gave way to a look of envy as Tim explained the methods and goals of IA, painting her a broad picture of the field. She was studying mechanical engineering at Michigan Tech and had never considered that archaeologists would study things like constructed mechanical systems in industrial process or workplaces. Her reaction raised many questions for us.

Over the years, precious few or no undergraduate engineering students enrolled in our field schools. This is odd, considering the appeal of IA to students of engineering and the importance Michigan Tech places on mechanical and industrial engineering; metallurgy; materials; engineering technologies; and environmental, civil, geological, and mining engineering and sciences. This is undoubtedly a result of the lock-step curriculum assigned for prospective engineers. Undergraduate engineering students at MTU are pressured to get summer job placements, co-ops, and internships that keep them on their tightly defined career track (see also Chap. 6).

If they are not in a co-op, students often spend summers redoing courses from which they withdrew during the year, trying to "catch up" with their cohort. Certainly, many undergraduate engineering students must also work summer jobs to earn money to pay for their studies the following year, and like a co-op or internship, this means they cannot go "into the field" for 6, 8, or 14 weeks away from campus.

So who are our field school students? From what groups do we draw people into our learning community? Like most archaeology research teams, Michigan Tech's are composed of people from many different backgrounds. While we hesitate to label individuals with demographic categories or to ascribe identities to them, over the years we have noticed that our field schools attract students and volunteers from varied life-stages. Our undergraduate student community at Michigan Tech is largely non-Hispanic White ($81 \pm 4\%$) and male (76%), reflecting general social patterns in STEM education within the United States (Scarlett 2007). As we mentioned above, however, Michigan Tech undergraduate students are usually in the minority on our research teams during field school. We also rarely see traditional undergraduate students from other universities, although those that do enroll are almost always studying history, anthropology, or archaeology, and rarely engineering.

The clear majority of field school enrollees stumble upon Michigan Tech and IA as a consequence of web searches. Others learn by "word-of-mouth" while traveling to see the artifacts, spaces, landscapes, or sites of industrial history. Very few enrollees find us using online databases like the *Archaeological Fieldwork Opportunities Bulletin*. Most of our field school enrollees are considering graduate studies in IA. About one half of our field school participants, however, are nontraditional students and they have usually worked for ten or more years before attending our field school. These practicing archaeologists, museum interpreters, photographers, artisans and craftspeople, engineers, and designers are so thrilled by their discovery of IA that they are often moved to apply to our graduate program. By joining the research team at field school, these students actually complete the first credits towards their graduate degree.

8.3   Foamers Are to Enthusiasts as Speed Freaks Are to Coffee Drinkers

At least one field school participant is often retired from a lifelong career, often in industry, technology, or engineering. These individuals choose to pursue their passion, studying industrial heritage. In 2009, Tim collaborated on a public archaeology project in Utah. One research team member used this simile to express the powerful passion that avocational researchers feel for the "big stuff" of industrial heritage. As archaeology has its fans that knap stone tools or replicate pottery, and history has its battlefield reenactors and buckskinners, so IA has individuals passionate about steam engines, locomotives and trains, foundries and furnaces, lathes, generators, grease monkeys, and machinery. In IA, many of these people often self-identify as "foamers," a nickname originally meant to be snide and pejorative, recalling a rabidly passionate enthusiast foaming at the mouth with excitement, standing in rapture before a running Corliss-type horizontal beam engine (with a 14″ diameter piston, 36″ stroke, and a 13-ft diameter flywheel, of course). Academics and professionals belittled this passion, criticizing the avocational individuals' lack of enthusiasm for advancing knowledge beyond antiquarian indexing, such as publishing comprehensive catalogs of machine types, locomotive engines, or surviving canal boats. Avocational IA communities have co-opted this nickname as a badge of pride, as is often the case with subcultures, and continue to thrive.

Our annual field school research teams often include at least one nontraditional student that might identify themselves as a foamer. Sometimes this person is a retired mining engineer, machinist, industrial manager, media specialist, or an agency land manager that discovered IA on the internet. Like the nontraditional students that enter our graduate program, these individuals join our research team because they are passionate about Industrial Heritage. These volunteers come to field projects through Elder Hostel, Earthwatch, or are simply interested in earning undergraduate credit. We find these people to be a tremendous asset to our learning environment during field school. They connect us to our intellectual history, since IA originally developed in alliance with adult continuing education programs in the United Kingdom (Cossons 2007: 12–16; Buchanan 2000, 2005) as well as the United States (Martin 2009: 286–289). In addition, IA has an advantage over general archaeology, as that field has systematically alienated most of its avocational communities. As we professionalized during the last century, and particularly during the past 40 years, scientific archaeology exerted a primary right to study antiquities using rigorous technical methodologies within defined ethical boundaries. As a consequence of that process, many professional and academic archaeologists took a moral high ground and drove others from the field, including museum curators, antiquarians, treasure hunters, looters, collectors, and all manner of hobbyists. Concurrent and parallel to that trend, archaeologists had long cultivated strained relationships with indigenous and aboriginal communities (Thomas 2000; Killion 2008; Ashmore et al. 2010). In the United States, this played out within the broader "culture wars" of the last 25 years.

While the archaeological heritage benefited from greater care and protection, and the discipline experienced great intellectual advances, archaeologists have recently spent a great deal of time struggling with the consequences of this alienation. Most particularly, professionals are trying to mend relations with aboriginal and descent communities. IA did not go through this. The avocational community still plays an important role in IA, attending annual national meetings of the Society for Industrial Archaeology and its local chapter events, taking factory and plant tours, interacting with students, and talking about their own passionate research. This is also true in the UK, and we would do well to remember that we industrial archaeologists didn't invent industrial tourism. America's elite traveled to see the early republic's industrial splendors along with its natural wonders (Gassan 2002), more than a century and a half before IA ever existed!

Foamers are also often stakeholders in industrial heritage. Many people with a passion for industrial history, architecture, or machines often work as engineers and mechanics in industry today. Many are deeply shaken and saddened by the structural adjustments of the American economy, particularly following the major changes during the last 10 years. Most are advocates for preservation in their own communities, where they have watched deindustrialization undo their life's work.

8.4 You Mean You Didn't Have Metal Shop in Junior High?

We increasingly come to rely upon foamers and other avocational and nontraditional students as key members of our field school research teams because many of them have direct experience with industrial labor. Fewer and fewer American university students have any experience with artisanal work, training as mechnicians,[5] or factory work generally. Over the past 50 years, most schools have shifted their academic programs to prepare students for postsecondary education, leaving little room for vocational-type classes. This has been exacerbated by "No Child Left Behind" educational policies, where schools now structure students' learning environment around testable, assessment-driven learning outcomes. Many school systems no longer require college-bound students to take wood shop, metal shop, mechanical or architectural drafting, home economics, or other experiential learning-based courses.

Fewer and fewer middle class undergraduate students in American universities are prepared to understand industrial labor. While some grew up gardening for example, few have had an opportunity to forge-weld using hammer and anvil. When they work, most undergraduates in the United States take jobs in service or retail industries, a trend that reflects ongoing structural changes in the American economy. Foundry and factory work are therefore as alien to most students as plantation or farm work. Our nontraditional field school students, who have returned to college after working in a steel mill, fishery, auto plant, rail yard, or mine, have consistently stepped into the role of peer-mentor, helping the younger students develop understandings of industrial work and labor.

The chronic disconnect between contemporary students and industrial activity is in part a generational experience that is increasingly becoming a population-wide phenomenon in the United States, as industries continue to relocate outside our communities. This movement of course is both symptomatic and symbolic of the changing and evolving nature of a capitalist world economy that in the modern era has informed and continues to inform the working lives of countless individuals globally. It is this connection, between the past industry of IA landscapes and the continuing cycle of industrialization, that we feel brings a particular relevance to the study of industry and labor.

---

[5] A mechanician is a practitioner of applied mechanics. In the twentieth century, professional engineers used this term to refer to anyone working with engineering mechanics. In the eighteenth and nineteenth centuries, the term would apply to any person working with practical applications of mechanics, trying to use physical theory to derive useful solutions for specific technological devices or systems.

8.5 5,000 Bricks Per Person Per Day?

Unfortunately, most students do not fully realize that we still live in an industrial economy. The markets of capitalism mystify the commodity chains that provide consumer goods to most Americans leaving only a vague sense of the processes that connect producers and consumers across disparate geographic regions and cultural conditions. Making students aware of the links that connect people across this global system problematizes these connections, but it is the practical exercise of doing IA that actualizes and personalizes these bonds.

We were both able to start teaching field schools in IA while teaching a class for academically gifted young students. Johns Hopkins University's Center for Academically Talented Youth allowed Tim to spend 10 years teaching a university-level introduction to archaeological sciences, with Sam serving as a teaching assistant for one of those years. As a part of that 3-week course, the 12- to 16-year-old students spent 1 week in the field recording the landscape and features at the site of the Lancaster Brick Works (1919–1979). We used the former brickyard as an outdoor classroom for our experimental archaeology labs as well as our fieldwork, and the students spent a lot of time clearing brush and moving piles brick – sometimes lots and lots of brick. The physical labor made a significant contribution to the learning environment because it created a sensory link to the industrial landscape.

As part of this course, the teenagers were able to learn from former brick workers. They met the last company president and interviewed other workers in the community. They heard the stories about how the managers supplemented regular staff by hiring hobos off the railroad to work in the yard. These men would assemble or unpack the kilns, a process during which one individual tossed two or more bricks per throw, pulling from a cart or pallet and feeding another man that was placing them as the kiln took form. Each individual threw at least 5,000 bricks per day at the Lancaster Brick Works. The students heard how the hobos and regular workers drank alcohol to numb the pain of the work. They heard testimony that despite the hard labor that bloodied people's hands, homeless people riding the rail knew that the Lancaster Brick Works yard was a good place to earn some money. The hot kilns of "Tickville" made a good place to camp on a cold winter's night after a day of work, only a short hike from an urban area, but also largely beyond the gaze of urban society.

Over time, we have come to believe that this physical labor is a critical part of the holistic learning environment during a field school because it opens industrial history to individuals who relate to the world through emotional and bodily kinesthetic intelligences. For many students, for example, this physical labor connects them to their own industrial heritage in their families. At Michigan Tech, we have been lucky to collaborate with students struggling with physical disabilities, elderly persons, and others that wanted to excavate and haul rubble as a regular part of a research team. We have always found creative solutions to these challenges.

8.6 I Just Can't See the People!

In designing our field schools, we often collaborate with modern artisans and include tours of operating industrial facilities when possible. Casting molten aluminum into hand-packed sand molds at Newburgh's Super Square Foundry[6] helped students understand the industrial processes that people performed in excavation areas at the West Point Foundry. Owner Dean Andersen and journeyman Amy Lahey believe passionately in the power of experiential learning and they shared their knowledge of craftsmanship with our students every year, helping them to learn to "see" some of the skills that make foundry work possible. Dean and Amy were interviewed by journalist (and fireboat engineer) Jessica DuLong in 2003 and she summarized their perspectives on increasing invisibility of hand-labor (2009: 237–240). Gordon and Malone (1994: 38–42) argued that artisanal skills and knowledge systems that accompany activities like patternmaking and sand molding are one of three knowledge or skill sets essential to understanding industrial production: work and artisanal skills, engineering and scientific skills, and organizational and management skills (Fig. 8.2).

In the first few weeks of the field school, students occasionally repeat critiques they have heard from previous professors, claiming that they "just can't see" the people we are studying. The students who express this have stumbled into 30-year-old stereotypes about the field or they have read work that pressures archaeologists to adopt a single unified research paradigm. This "invisible worker" critique arose as part of a larger indictment of archaeologists' tendency to treat subjects of study as "faceless blobs" (Tringham 1991: 94). Ruth Tringham's famous faceless blobs dovetailed nicely with the industrial age's anonymous proletarian masses of de- and unskilled laborers. Critics looked at detailed analyses of machines or schematic drawings of industrial processes produced in HABS/HAER surveys, and the only people they "saw" were the tiny figures included for scale in isometric drawings that illustrated the studies. They rightly faulted scholars of IA for continuing to write "big man history" that focused upon the inventors, engineers, capitalists, and political leaders that built large and complex technological systems, ignoring the contributions of mute workers or the social negotiation of work. Today, industrial archaeologists regularly turn their attention to the "plurality of power" in industrial capitalism and its communities (Cowie 2011; Shackel 1996).

The routine act of doing IA can illuminate the presence of these past "invisible workers." Students excavating in a corner of the molding shop ruins at the historic West Point Foundry uncovered a molder's shovel embedded in a pile of sand. The shovel had been abandoned by a foundry worker as he walked away from the pile of

---

[6] Super Square Ironworks: 545 Broadway, Newburgh, NY 12550, USA; mail address: Super Square Corporation, PO Box 636, Beacon, NY 12508, USA (845) 565-3539.

Fig. 8.2 Michigan Tech industrial archaeology students, both undergraduate and graduate, draw upon experiential learning activities, such as making these molds for casting aluminum at Super Square Foundry in 2006: (**a**) Lindsay Kiefer coats a wooden pattern and mold with parting compound as Amy Lahey prepares molding sand; (**b**) Lindsay sifts sand into the mold before packing it around the pattern; (**c**) Stephanie Atwood removes the pattern from the packed mold; (**d**) Dean Andersen pours metal into the molds (photos courtesy of Michigan Technological University)

molding sand he had been shoveling nearly 100 years earlier. Serendipitous moments like these in the field, and others much more mundane, forge an affinity between students and the daily lived existence workers in the industrial past. Field students who have been heaving shovels full of dirt themselves for weeks begin to make the connection between their labors and those of former workers at the site. In this way, the sensory connection to place shared through the act of physical labor again embodies the industrial landscape.

We design our field school to blend experiential learning opportunities with traditional archaeological training of skills like mapping, drawing, and excavating. When combined with the peer mentoring from the nontraditional students in our learning community, and to a lesser extent the emotional learning from the physical labor, the field school creates a powerful combination that resists Tringham's faceless blobs. Students begin to understand different ways to "see" the individuals that inhabited a workplace without theorizing them into predetermined boxes in order to understand them. Our field schools almost always involve taking the students out of the industrial core and considering the connections of the workplace and the forces of production to surrounding communities, households, families, and landscapes, linking production and reproduction and the local with the global.

Both Beaudry (2005) and Hardesty (2000) have advocated for bringing multiple voices into IA. Multivocality should not only include focusing academic attention on how people negotiated social power, gender, ethnicity, and identity when they worked in industrial jobs. A truly multivocal IA also values the multiple ways in which people relate to and understand those stories and experiences. A "postprocessual" or "processual plus" archaeology should value different ways of "seeing into" the industrial past.

We try hard to get students to put aside seductive academic debate and first explore the different ways of relating to the material residues of industry. They learn to "see" a blast furnace or steam engine through the eyes of a foamer; see work process through an engineer's eyes; recognize artisanship from things, as one craftsperson can do using the work of another; and struggle with monotonous days thrusting wheelbarrows of brick over rough ground. Then they can "see" the critiques and concerns of social reformers, environmentalists, capitalists, critics, progressives, theorists and artists, and identify the social, economic, and political divisions and unities that may have existed within a landscape of work beyond simple binary distinctions between labor and capital. The most interesting research occurs at the intersections of those various ways of seeing. We believe these ways of seeing will help young professionals develop meaningful relationships with members of descent communities and other stakeholders in the heritage they study or manage.

8.7 Resistance! Resistance!

There is no doubt that IA thrusts students directly into contentious areas of American culture, particularly perspectives on work. We admire our colleagues who position their research in these contentious areas, working collaboratively to produce new knowledge about industrial heritage sites or industrial societies. Exemplary projects include the Colorado Coal Field War by the Ludlow Collective (McGuire and

Reckner 2003; Walker and Saitta 2002), The Levi Jordan Plantation Project (McDavid 2004), and the collaborative archaeology of homeless communities (Zimmerman et al. 2010). The archaeologists working on these projects have very different perspectives and generally would not consider their research to be part of IA. We are working with our students to generate new understandings of the industrial world however, not reinforcing traditional disciplinary boundaries.

During one of the early seasons studying West Point Foundry, periodic shouts drifted through the trees, "Hey... Do you know what this is? Resistance!!!" Surveyors occasionally called to the rest of the team, reporting a newly discovered broken beer bottle or parts of a stolen shopping cart. At the time, the giggling was perplexing. Only later did we realize that students were teasing each other about a particular archaeological report in which the author had identified recovered artifacts as material residue of workers' resistance to management control. In the learning environment of the field school, this mix of students – undergraduate and graduate, traditional and nontraditional – had developed a collaborative critique. While their individual interests varied regarding the hidden transcripts of resistance, they had decided as a group that they did not like the simplistic way this particular author (or authors) had linked recovered objects with the power relationships in an industrial community and workplace. In a complex, social working environment like the West Point Foundry, the field team had decided that such a monolithic view of capital and labor seemed hopelessly naïve (Fig. 8.3).

Field school learning is situated learning. The best-designed research project serves multiple stakeholder communities, with specific care to collaborate with members of communities underrepresented in university life.[7] Most American

---

[7] The success of the "Dig Where You Stand" and "study circle" movements in Sweden (and related programs in Denmark, Norway, and Finland) is still little known in the United States as models for public archaeology in industrial communities. The "Nordic Tradition" has old roots in the region (Burchardt and Andresen 1980:25–29). Between 1945 and 1970, Folklore and Oral History programs involved tens of thousands of Scandinavians in documenting the transformation of life consequent to industrialization. Following Gunnar Sillén's publication of *Stiga vi mot ljuset: Om documentation av indusrti- och arbetarminnen* [Towards the light we ascent: On documentation of industry and workers' memories] in 1977 and Sven Lindqvist's publication of *Gräv där du Står: Hur man utforskar ett jobb* [Dig where you stand: How to explore a job] in 1978, a popular and widespread movement arose which involved collectives of industrial workers who collaborated with "Working Life" Museums and The Workers' Educational Association (collaboratively run by the Swedish Social Democratic Party and several trade unions). By 1984, there were more than 1,000 community study groups in Sweden involved in archaeological, historical, genealogical, and oral history research, writing factory histories, biographies of industrial workers, and social histories of their own communities. Anna Storm estimated that between 10,000 and 100,000 people were inspired to this movement because it transformed regular people into *creators* of heritage, rather than *consumers* of cultural history, performances, documentation projects, or interpretations produced by intellectuals and professors (Storm 2008:39–43). By comparison, Cossons (2007:13) noted our current lack of academic insight into the motivation of avocational industrial archaeologists that set up local IA organizations throughout UK in the 1950s and 1960s, but recalled that the Workers' Educational Association had played an important role (cf. Speight 1998, 2004).

Fig. 8.3 The annual field school focuses upon many different types and scales of industrial work: (**a**) students excavating a kiln at a family-operated pottery in rural Parowan, Iron County, Utah, in 2009; (**b**) a research team excavates the enormous cupola furnace base in the Casting House Complex at the West Point Foundry, Cold Spring, New York, in 2007 (photos courtesy of Michigan Technological University)

industrial communities include people of widely varied backgrounds with dramatically different ideas about work, labor, and the relations of production, like our field school research teams. Some individuals hate any idea of corporate paternalism. Others believe that communities can and should be designed to mitigate the hazards of industrial life. Many believe that direct collective action is the best method for improving one's living conditions. An equal proportion believes strongly in the ennobling power of work, viewing work as still fundamental to Americans' self-identity and collective thoughts about society. Despite these deeply held and conflicting ideas, the field school produces new knowledge about industrial society through constructive collaborations. Because archaeological fieldwork contains inescapable and essential ambiguities, these varied people work together (and usually comfortably) to reconcile their perspectives on understanding what we are learning. This happens despite the crushing rhetoric of modern social discourse encountered in 24-7 cable TV punditry and the incessant vitriolic spew of internet discussions. We think this happens for several reasons. Field school labor is intensive, as we mentioned, and it is authored. One's journal and paperwork enter the permanent research archive. The work also takes time. Without the anonymity and brevity provided by modern media and styles of discourse, people are generally civil and constructive, even during passionate disagreements.

We almost always undertake our field schools as public archeology. In addition to confronting issues of class and identity politics among themselves, students and team members constantly find themselves negotiating public tours of industrial heritage sites. Visitors come to see our archaeological digs at the foundry, mine, mill, smelter, fishery, pottery, or wherever, and they bring their ideas: anger over environmental degradation or over environmental regulations; beliefs in the ennobling or emasculating power of work; blame directed at labor unions or Wall Street investors and multinational corporations for ruining domestic industry; a sublime or romantic attachment to the scale or landscapes of industrial production; hatred or love for globalization; or convictions about the perceived evils or benevolence of corporate paternalism, religious institutions, or company towns.

We challenge students to engage with people from these different perspectives, meeting them respectfully as equals. We also model these attitudes ourselves, demonstrating the value of different intellectual perspectives. We discuss research themes and field methods from the social sciences, humanities, engineering, and design. We value the different perspectives of our colleagues, including those building generalized patterns of human behavior, weaving micro-historical or biographical narratives, applying frameworks from evolutionary biology, positioning an activist scholarship of political economy, or studying the social construction and evolution of technological systems. Echoing the thoughts of Ronald Reno in his study of charcoal burners in Nevada's Eureka mining district, when "[t]aken together, this diversity of approaches and sources produce[s] a historical ethnography of a functioning industrial culture" (Reno 1996: 317). Similar to the functioning of industrial cultures of the past, students come to realize that industrial landscapes today, like yesterday, are more about negotiation than resistance.

8.8   "Don't Trip on the Mining Machinery While Enjoying the Virgin Splendor of This Wilderness!" Or "…and Then the Test Trench Groundwater Dissolved the Styrofoam Coffee Cup!"

IA also puts field school students at the center of cultural debates about industrial production and environmental sustainability. Industrial heritage complicates often-easy alliances between heritage preservation and environmental restoration or open-space movements. These tensions are perfectly captured in The Michigan State Historic Site marker along the road into the Porcupine Mountains State Park. The marker reads in part, "Machinery, rock dumps, and old adits are ghostly reminders of forty mining ventures in the years from 1846 to 1928…. Some logging took place around 1916…. Finally in 1945 the area was made a state park to preserve its virgin splendor." The students in our most recent field trip found this paradoxical marker hysterical, as they trudged into the woods to see this virgin (that is unsullied, unspoiled, modest, and initial) example of industry in the woods of far northern Michigan.

Students usually come to our field school with a simplistic notion of "industry vs. environment." That industry despoils nature has been a widely held belief in American society, a belief that has deep roots in western intellectual tradition (Glacken 1967) and took its current form following the birth of the modern environmental movement (Carson 1962). All productive activities leave communities with ecological legacies, economic challenges, and social problems.

Ultimately, all industrial heritage sites represent failures. While some factories operated longer than others, or perhaps one mine returned more on investment than another, all industrial operations eventually end. The natural resources are extracted and what remains cannot be profitably won on an industrial scale. Manufacture eventually becomes too expensive, facilities outdated, and capital flees to cheaper markets. IA often brings research teams to "brownfields," "Superfund sites," and other degraded and contaminated landscapes that by no stretch of the imagination can be considered "virgin," yet contain great potential to yield material evidence of human industrial activity (Quivik 2000, 2007; Symonds 2004, 2006; White 2006).

Many of these sites and landscapes pose serious threat to people's health. We tell our students a story about IA and urban-sites archaeology in which the Styrofoam cup serves as the punch line about the hazards of doing archaeology in urban and industrial settings. In this archetypal story, a colleague working in the backhoe trench began to develop a headache and noticed a funny smell. The crew chief passed down an empty coffee cup for the person to scoop up a groundwater sample that they could later have analyzed. In a matter of seconds, chemicals in the water dissolved the Styrofoam cup. Everyone immediately scrambled out of the excavation and work came to a halt as the team realized they were facing a potential medical emergency.

Unfortunately, this story is neither allegorical nor is it exaggerated; rather this cautionary tale and others like it serve to warn IA students away from a cavalier "cowboys of science" mentality that can be found in both general archaeology and IA. We think that English archaeologists led the way addressing health and safety concerns, when the Council for British Archaeology published a pamphlet explaining legally required safety requirements (Fowler 1972). Through the 1990s in the United States, a growing list of professional publications drew attention to the heath hazards of both field- and museum-based studies involving archaeological (McCarthy 1994; Flannigan 1995; Poirier and Feder 2001), forensic (Fink 1996; Walsh-Haney et al. 2008), and ethnographic or natural history collections (Odegaard and Sadongei 2005). In the United States, the caviler archaeological mentality began to wane as professional practice developed largely within the Society of Professional Archaeology, particularly in their publication, the *SOPA Newsletter* (cf. Murdock 1992; Garrow 1993; Fink and Engelthaler 1996) and *Federal Archaeology* (cf. Flannigan 1995). This trend culminated in the publication of *Dangerous Places: Heath, Safety, and Archaeology* (Poirier and Feder 2001). Safe and professional practices have begun to percolate into introductory field manuals to varying degrees.[8]

All archaeology conveys risks to health and safety: confined spaces excavation, pathogens and occupational diseases, unstable historic architecture, temperature stress, sharp tools, toxic plants and venomous animals, and even the crew's social practices are all concerns (Langley and Abbott 2000). By its very nature, however, IA will more often bring professional, student, and avocational practitioners into contact with hazardous threats. One half of *Dangerous Places* examines hazards posed by colonial and industrial activity (of particular note are Hatheway 2001; Roberts 2001; Saunders and Chandler 2001; Reno et al. 2001). Industrial processes like tanning leather, making paper, dyeing textile, extracting metals for ore, and founding steel all involve chemicals like amyl acetate, sulfuric and other acids, hydrogen chloride, benzene, naptha phenol, toluene, and elements such as lead,

---

[8] Typical examples of health and safety concerns addressed in these books include *brief* mentions of regulations regarding excavations in deep trenches (Black and Jolly 2003:61, 64–65; Carmichael et al. 2003:52; Purdy 1996:96); recommendation to get a tetanus booster and pay up on your insurance policy (McMillon 1991); a discussion of disease risk and prevention, proper tool use, hygiene, and a paragraph about deep trenches, Occupational Safety and Health Administration (OSHA) standards, state safety checklists, and legal liability waiver forms (Hester et al. 1997:110–112); discussions of employee safety training, regulations and shoring regarding deep excavations, cold temperatures, and working in the woods during hunting season (Neumann and Sanford 2001:68, 160–161, 186–189); and emergency first aid and strategies for dealing with disaster (Kipfer 2007:171–179, 193, 212). British and Australian archaeologists have done a much better job including careful discussions of safety and health issues, and we point to Roskams's (2001:82–92) extensive discussion of issues in a dedicated section of his manual, but also point to the fact that he has also made themes of safe and careful professional practice a regular part of the narrative throughout the book. Heather Burke and Claire Smith, along with Larry Zimmerman, also included extensive discussion about health and safety issues in their field handbooks (Burke and Smith 2004; Burke Smith and Zimmerman 2007:134, 194–196; Smith and Burke 2007:96–108, 117–123). This last set of books also hints that field manuals with discussions of Industrial Archaeology and Urban Archaeology among the spectrum of archaeological practice give more serious thought to health and safety policy and practice (along with those directed toward students seeking to become Cultural Resources Management professionals).

arsenic, mercury, chlorine, and chromium. We deal with so much rusted iron that we strongly recommend TETANUS vaccinations for all team members and we occasionally had discussions about unexploded ordinance (UXO) while at the West Point Foundry; fortunately however, we have not lead a field crew into a highly contaminated site. Team leaders should research and anticipate health and safety risks posed by each new project. This should be part of their preparations for the study, often in collaboration with environmental scientists and public health professionals. Many government health services and NGOs also provide ready access to information about occupational health.[9]

As a department, we created the Ph.D. in Industrial Heritage and Archaeology, in part, to establish closer ties between the academic study of industrial heritage sites and social and environmental consequences of industrial wealth production. Industrial activities transformed (and continue to transform) the world as never before in the human experience. While our students might study a particular industrial site or community, they also face the living community's struggles with the consequences of producing industrial wealth in a capitalist world. Heritage preservation seems to be a great idea, and archaeological heritage easily links with intangible cultural heritage and environmental heritage conservation, until effluent from a heritage site is linked to cancer in children living downstream. Those same youngsters, however, live as part of an industrial community with rich and textured relations to their heritage sites and landscapes, as does any other stakeholder group or decent community with any other type of heritage. "Hard places" and landscapes, as Robertson (2006) wrote, often become enduring expressions of shared physical work, risk, and sacrifice that are important to family and community.

Individual students on Michigan Tech's IA Field Teams are forced, along with the project as a collective, to reconcile the fact that academic research is performed in the contemporary world. Creating new knowledge includes social and political outcomes beyond academic research questions. Students are shocked to find that some community stakeholders see them as neocolonial tools of the wealthy, urban, and educated elite that employ environmental or historic preservation laws to preserve quaint, picturesque landscapes for vacation, while other community members are happily bending the field school process to meet their own private political or social objectives. The subtleties and complexities of these social negotiations are normal in IA, and projects must often struggle to reconcile advocacy for environment and advocacy for various descendent-, local-, and other stakeholder communities (McGuire with the Ludlow Collective 2008: 216–217).

---

[9] Examples of these resources include The United States Department of Labor's OSHA publication of standards and guidelines for excavation as well as standardized format guidelines for Material Safety Data Sheets (MSDS) for chemicals. The MSDS format includes information on handling and storage, toxicity, fire risk, and first aid procedures and has been widely adopted by other government and NGO groups, such as the provincial health services of Canada (http://msds.ohsah.bc.ca/). The European Agency for Safety and Health at Work (EU-OSHA) and the European Chemicals Agency (ECHA) compiled the standards and practices of member states, including details like the Globally Harmonized System for the Classification and Labeling of Chemicals (GHS).

Our field school research teams are constantly confronted with the question of what is an authentic landscape and how do changing perspectives reflect changing attitudes towards industry. In other words, is an industrial landscape nature despoiled, a landscape of transformation and progress, or something else entirely? As indicated above, most industrial archaeologists understand waste as fundamental to production and therefore wastes are important sources of information about an industrial site. Over the course of the field school, students start to understand the complexities of social constructions like sustainability, toxicity, risk, and heritage and they appreciate the challenges confronting communities trying to make decisions about these sites (cf. Gorman 2001). A community may be proud of its industrial heritage, for example, and some members may advocate for preserving it, but at the same time state environmental officials might require that the industrial landscape be "mitigated" for toxic materials, potentially erasing all traces of past industry.

In Michigan Tech's recent study of the Cliff Mine in Keweenaw County, Michigan, the field teams had to explain to visitors that the United States government's Environmental Protection Agency and the Michigan State Department of Environmental Quality had both determined that stamp sands were leaching metals into Eagle River, contributing to environmental contamination in its watershed. Those agencies required the sands be removed or encapsulated. At the same time, local newspapers printed a press release from our own university which reported that both those agencies had also determined the same stamp sands to be "safe for full body exposure" and approved permits allowing them to be used for the manufacture of asphalt shingles for domestic homes (Gagnon 2010). Residents, descendents, other stakeholders, and students often find these actions contradictory, incomprehensible, and ultimately frustrating. Our field school participants realize that they are doing much more than discovering new knowledge about the industrial past. They are often negotiators or facilitators, helping individuals and various communities of stakeholders navigate these difficult and emotional issues. It is through exposure to this process in the field that students begin to recognize the complexities of balancing questions of environmental, cultural, and economic sustainability as part of IA projects and industrial heritage management.

An increasing number of industrial archaeologists call for research to be centered back in the real world, confronting and engaging social conflicts surrounding the clean-up of waste and the management of existing abandoned industrial structures.[10] We design our field school experiences to put students into situations like these, which require students to help generate new knowledge for academic discussion about industrial history on projects that will also have useful and relevant outcomes

---

[10] These calls come from both Industrial and Historical Archaeologists concerning environmental remediation, ecological or economic justice (Joshi 2008; McGuire and Reckner 2005; White 2006), economic redevelopment, and cultural revitalization and education (de Haan 2008; Dong 2008; Greenfield and Malone 2000; cf. Gross 2001; cf. Palmer 2000). These issues became increasingly clear as *Industrial Archaeology* grew into *Industrial Heritage* and is therefore increasingly tied to the powerful "design culture" that surrounds adaptive reuse, sustainable redevelopment, and tourism (Conlin and Jolliffe 2011; Hamm and Gräwe 2010).

in the real world. Doing this fieldwork often induces cognitive dissonance in students, faculty, and other research team members. We must all deal with conflicting and seemingly irreconcilable points of view between academic paradigms, real world priorities, and situations such as the "ecological" land agencies that manage industrial heritage landscapes.

8.9 I'm Not Really Doing *Dirt* Archaeology

It is not surprising that dissonance characterizes contemporary perceptions of the broader meaning and value of past industry, industrial labor, and industrial landscapes. The Modern Era, which has been dominated by a global capitalism predicated on increasing industrial production, entangling networks of distribution, and discrepant patterns of consumption, is rife with incongruities and inequities that have become naturalized as part of modern life. As researchers and students studying the historical period, we benefit from a multiplicity of data sources that help to explore the potential meaning and relevancy of industry to both past actors and present participants.

As instructors, we actively introduce students to a broad spectrum of methodological and theoretical approaches from both within the field of archaeology and from other disciplines in the Social Sciences that encourage a multivocal IA. Traditionally, this process of exposure begins with the requirement that all incoming graduate students participate in the annual summer field school. The field school is ideally intended to serve as an initial exposure to a broad IA approach that emphasizes the variety of data sources from which scholarship can grow, e.g., material culture, written records, photographs and photography, architecture and the built environment, oral history, landscape studies, and environmental data, and that these approaches are all part of a multidisciplinary *archaeological* approach. However, summer invariably ends and the realities of the academic year set in.

It is not uncommon in the first weeks or months of a student's tenure in the department to hear some of them dogmatically state "I'm not really doing *dirt* archaeology," referring to the long-standing orthodoxy between IA communities, including a history of technology community centered on machines, buildings, and technological processes; an ethnographic or social history community focused on oral history and testimony; and the community in generalizing historical archaeology that unearths social meaning by moving dirt. In our students we are at once confronted with the historical legacy of a bounded IA established in the study of technological system builders and "their" workers, as a study independent from archaeological investigation. Our students gravitate to one professor or another, hitching their careers to one funded project or another, targeting jobs with agencies, companies, or future academic departments. The students tend to surrender the holistic and interdisciplinary view of archaeology.

Some students are fascinated by current academic debates in which some scholars wish to refocus IA on the social experience of industry and the negotiation of community or the identity politics of consumerism and consumption (cf. Casella and Symonds 2005). Others feel a powerful romantic attachment to industrial ruins, like so many international artists drawn to the picturesque decay of abandoned industrial facilities and the poetic "purposelessness of places of work stranded by abandonment" (Cossons 2007: 18). A few strive to understand a particular type of technology or sector of industrial production. Many feel increasing urgency as with the sudden shifts brought about by the current "Great Recession," seeking to help industrial communities with development while preserving tangible remains.

As younger scholars in one of the leading programs in this field, we embrace the necessity of positioning IA as a research endeavor that emphasizes the multiple voices of the past and the importance of this past in a multivocal present. However, while we enthusiastically broaden the perspective of IA to include the social dynamics of industrial life, we should not abandon the established strengths of the field, including an interest in the history of technology and the social construction of technological systems. An IA that combines the compelling systems-oriented thinking of contemporary social theories with the insight of narrative-based historical studies of individuals or technological systems can help to reduce the mystification and alienation that surrounds the functioning of the economic world-system.

As our students fall into the trap of traditional or emerging academic and bureaucratic niches, we encourage them to continue the multivocal thinking from field school. "Try explaining the complexities of the current global financial crisis," we tell them, "without moving between structural explanations of financial systems; the 'big man' style biographies of people who engineered, facilitated, or managed the collapse; and the individual narratives of people who's lives were transformed by it." These are all essential tools and perspectives if one is to understand the story of life in an industrial, capitalist world.

8.10   Conclusions

After years of directing IA field schools, we have become convinced that we should encourage students to approach the industrial past as multilayered landscapes. Upon these landscapes, we approach the physical and social environments of workplace, neighborhood, and community as products of the negotiation between local, regional, and global phenomenon and people. Documenting local processes enables students to demystify the "postindustrial world" and serves to reassert the fundamental connections between producers and consumers, both past and present, as participants in a capitalist world-system. In this sense, *both* the act of doing IA research in a place *and* the intellectual questions posed in IA both deconstruct the myth of a "postindustrial" world.

Developing field schools for a postcolonial IA will be one of our greatest future challenges. IA, and by extension, field schools in IA have the potential to further our understanding of the contemporary world by considering industry from alternate vantage points that move beyond the privileged perspective of Western industrial history. The IA of the future will need to view industrial history from the perspective of both the "core" and the "periphery." This will mean moving IA research into geographic regions, modes of production, and industries that have been traditionally outside the realm of IA studies. Moreover, these studies will need to explore the global ramifications of industrialization by elucidating the diverse ways in which variables such as race, ethnicity, gender, and class, along with processes such as colonization, globalization, and Westernization, came to increasingly structure people's lives under capitalism.

Most arguments over how to define IA are rooted in the basic question of who should "control" the study of material remains of industrial life, who should set the agenda by which we measure our success. We agree with Cossons (2007) that IA derives its intellectual vigor from its diverse participants, both applied and academic. The discovery of new knowledge about the industrial world, both topical and theoretical, must be linked to practical and tangible outcomes for descent and stakeholder communities.

In teaching our field schools, we do not try to insert a new master narrative to replace those that have come before, but instead seek to reinforce our existing connections and establish new voices in the discussion. We must also consider that field schools disadvantage certain groups of students. Students studying for engineering degrees, those from working class backgrounds, and nontraditional students all have obligations or commitments that prevent their participation in a 6-, 8-, or 12-week field school programs away from campus during the summer.

We must expand our existing collaborative learning projects, particularly by deemphasizing the exclusivity of remote field schools and undertaking more local archaeological fieldwork during the academic semester. The goal should be to create more inclusive field schools that integrate students as part of collaborative teams, working with people from many perspectives and institutions, in an environment that encourages both experiential and intellectual learning.

IA is a vibrant area of international scholarship driven by the conviction that the development and spread of industrial society is the most significant global transformation in human history. This research is also occurring amid the extraordinary deindustrialization of developed regions and the transformative development of other communities around the world. Ultimately, field schools in IA should create a multivocal atmosphere in which students can produce new knowledge while also tackling real world problems related to those experiences.

Acknowledgments The authors wish to thank Harold Mytum for the opportunity to contribute to this important volume. We are grateful to our colleagues and students in the Department of Social Sciences at Michigan Technological University and our other collaborators and supporters for all that they have contributed over the years. We would like to single out Patrick Martin, Elizabeth Norris, Steven Walton, Paul White, and Susan Martin for their contributions to 10 years of collaborative teaching of field schools at the West Point Foundry archaeological site, as well as all the many other research team members we cannot list here. The Scenic Hudson Land Trust supported our collaborative research during those years, and we are grateful for their commitment to archaeological study and public outreach. Sean Gohman has been a critical collaborator in our design of a public archaeology field program at the Cliff Mine site. The Cliff Mine project is supported by Heritage Grant from the Keweenaw National Park Advisory Commission and gifts from LSGI Technology Venture Fund L.P., Joseph and Vickey Dancy, Paul LaVanway, and Bill and Eloise Haller.

References

Alfrey, J., & Putnam, T. (1992). Industrial culture as heritage. In J. Alfrey & T. Putnam (Eds.), *the industrial heritage: Managing resources and uses* (pp. 1–39). London: Routledge.

Areces, M. Á. Á., & Tartarini, J. D. (Eds.). (2008). *Patrimonio Industrial en Iberoamérica: Testimonios de la memoria del trabajo y la producción.* Buenos Aires: Museo del Patrimonio de AySA/UNCUNA.

Ashmore, W., Lippert, D. T., & Mills, B. J. (Eds.). (2010). *Voices in American archaeology.* Washington: Society for American Archaeology.

Beaudry, M. (2005). Concluding comments: Revolutionizing industrial archaeology? In E. Casella & J. Symonds (Eds.), *Industrial archaeology: Future directions* (pp. 301–314). New York: Springer.

Black, S. L., & Jolly, K. (2003). *Archaeology by design.* The Archaeologists toolkit 1. Walnut Creek: AltaMira Press.

Buchanan, A. (2000). The origins of industrial archaeology. In N. Cossons (Ed.), *Perspectives on industrial archaeology* (pp. 18–38). London: Science Museum.

Buchanan, A. (2005). Industrial archaeology: Past, present, and prospective. *Industrial Archaeology Review, 28*(1), 19–21.

Burchardt, J., & Andresen, C. E. (1980). Oral history, people's history, and social change in Scandinavia. *Oral History, 8*(2), 25–29.

Burke, H., & Smith, C. (2004). *The archaeologist's field handbook.* Crows Nest: Allen & Unwin.

Burke Smith, H. C., & Zimmerman, L. J. (2007). *The archaeologist's field handbook.* Walnut Creek: AltaMira Press.

Carmichael, D. L., Lafferty, R. H., III, & Molyneaux, B. L. (2003). *Excavation.* The archaeologists toolkit 3. Walnut Creek: AltaMira Press.

Carson, R. (1962). *Silent spring.* Boston: Houghton Mifflin.

Casella, E. C. (2006). Transplanted technologies and rural relics: Australian industrial archaeology and questions that matter. *Australasian Historical Archaeology, 24,* 65–75.

Casella, E. C., & Symonds, J. (Eds.). (2005). *Industrial archaeology: Future directions.* New York: Springer.

Cerdà, M. (2008). *Arqueología Industrial: Teoría y Prática.* València: Universitat de València.

Conlin, M. V., & Jolliffe, J. (2011). *Mining heritage and tourism: A global synthesis.* New York: Routledge.

Cossons, N. (2007). Industrial archaeology: The challenge of the evidence. *The Antiquaries Journal, 87,* 1–52.

Cowie, S. (2011). *The plurality of power: An archaeology of industrial capitalism.* New York: Springer.

Crandall, W., Rowe, A., & Parnell, J. A. (2003). New frontiers in management research: The case for industrial archaeology. *The Coastal Business Journal, 2*(1), 45–60.

de Haan, D. (2008). Ironbridge institute heritage archaeology course. In J. af Geijerstam (Ed.), *The TICCIH seminar on training and education within the field of industrial heritage* (pp. 34–38). Resource document. TICCIH. Retrieved April 1, 2011, from http://www.mnactec.cat/ticcih/docs/1255330147_stockholmedsem.pdf

Dong, Y. (2008). The status and problems of the research on industrial heritage in China. In J. af Geijerstam (Eds.), *The TICCIH seminar on training and education within the field of industrial heritage* (pp. 39–41). Resource document. TICCIH. Retrieved April 1, 2011, from http://www.mnactec.cat/ticcih/docs/1255330147_stockholmedsem.pdf

DuLong, J. (2009). *My river chronicles: Rediscovering America on the Hudson.* New York: Free Press.

Ebert, W., & Bednorz, A. (1996). *Kathedralen der Arbeit: Historische Industriearchitektur in Deutschland.* Tübingen: Wasmuth Verlag.

Fink, T. M. (1996). Rodents, human remains, and North American Hantaviruses: Risk factors and prevention measures for forensic science personnel – A review. *Journal of Forensic Sciences, 41,* 1052–1056.

Fink, T. M., & Engelthaler, D. M. (1996). Health issues in Arizona archeology: Update 1995. *SOPA Newsletter, 20*(1/2).

Flannigan, J. (1995). What you don't know *can* hurt you. *Federal Archaeology, 8*(2), 10–13.

Fowler, P. (1972). *Responsibilities and safeguards in archaeological excavation*. London: Council for British Archaeology.

Gagnon, J. (2010). Alumnus teams with tech to reclaim stamp sand and grow an industry. *Michigan Tech News*. Electronic document. Retrieved September 15, 2010, from http://www.mtu.edu/news/stories/2010/august/story30658.html

Garrow, P. H. (1993). Ethics and contract archaeology. *SOPA Newsletter, 17*(9/10), 1–4.

Gassan, R. H. (2002). The birth of American tourism: New York, the Hudson Valley, and American culture, 1790–1835. PhD dissertation, *Electronic doctoral dissertations for UMass Amherst*. Paper AAI3056228. Retrieved September 27, 2007, from http://scholarworks.umass.edu/dissertations/AAI3056228

Glacken, C. J. (1967). *Traces on the Rhodian shore: Nature and culture in western thought from ancient times to the end of the eighteenth century*. Berkeley: University of California Press.

Gordon, R. B., & Malone, P. M. (1994). *The texture of industry: An archaeological view of the industrialization of North America*. New York: Oxford University Press.

Gorman, H. S. (2001). Conflicting goals: Superfund, risk assessment, and community participation in decision making. *Environmental Practice, 3*, 27–37.

Greenfield, B., & Malone, P. (2000). "Things" that work: The artifacts of industrialization. *OAH Magazine of History, 15*(1), 14–18.

Gross, L. (2001). Industrial archaeology: An aggressive agenda. *IA: The Journal of the Society for Industrial Archaeology, 27*(1), 37–40.

Hamm, O., & Gräwe, C. (2010). *Bergbau Folge landschaft/post mining landscapes*. Berlin: Jovis-Verlag.

Hardesty, D. L. (2000). Speaking in tongues: The multiple voices of fieldwork in industrial archaeology. *IA: The Journal of the Society for Industrial Archaeology, 26*(2), 43–47.

Hatheway, A. W. (2001). Former manufactured gas plants and other coal-tar industrial sites. In D. A. Poirier & K. L. Feder (Eds.), *Dangerous places: Health safety, and archaeology* (pp. 137–156). Westport: Bergin & Garvey.

Hester, T. R., Shafer, H. J., & Feder, K. L. (1997). *Field methods in archaeology* (7th ed.). Mountain View: Mayfield Publishing Company.

Joshi, M. (2008). The case for salvaging the remains of the world's worst industrial disaster as Bhopal, India. *TICCHI Bulletin, 43*(winter), 1, 8.

Killion, T. W. (2008). *Opening archaeology: Repatriation's impact on contemporary research and practice*. Santa Fe: School for Advanced Research.

Kipfer, B. A. (2007). *The archaeologist's fieldwork companion*. Malden: Blackwell Publishers.

Komatsu, Y. (1980). Industrial heritage of Japan. *Industrial Archaeology Review, 4*(3), 232–234.

Langley, R. L., & Abbott, L. E., Jr. (2000). Health and safety issues in archaeology: Are archaeologists at risk? *North Carolina Archaeology, 49*, 23–42.

Lindqvist, S. (1978). *Gräv där de Står: hur man utforskar ett jobb*. Stockholm: Bonnier.

Martin, P. (1998). Industrial archaeology and historic mining studies at Michigan Tech. *CRM Magazine, 21*(7), 4–7.

Martin, P. (2001). The importance of networking and the American IA experience. In M. Clarke (Ed.), *Industriekultur und Technikgeschichte in Nordrhein-Westfalen, Initiativen und Vereine* (pp. 107–110). Essen: Klartext-Verlag.

Martin, P. (2009). Industrial archaeology. In T. Majewski & D. Gaimster (Eds.), *International handbook of historical archaeology* (pp. 285–297). New York: Springer.

McCarthy, J. P. (1994). Archaeologists in Tyvek: A primer on archeology and hazardous materials environments. *SOPA Newsletter, 18*(2), 1–3.

McDavid, C. (2004). From "traditional" archaeology to public archaeology to community action: The Levi Jordan plantation project. In P. A. Shackel & E. Chambers (Eds.), *Places in mind: Public archaeology as applied anthropology* (pp. 35–56). New York: Routledge.

McGuire, R. H., & Reckner, P. (2003). Building a working class archaeology: The Colorado Coal Field War Project. *Industrial Archaeology Review, 25*(2), 83–95.

McGuire, R. H., & Reckner, P. (2005). Building a working class archaeology: The Colorado Coal Fields War Project. In E. C. Casella & J. Symonds (Eds.), *Industrial archaeology: Future directions* (pp. 217–241). New York: Springer.

McGuire, R., & The Ludlow Collective. (2008). Ludlow. In R. McGuire (Ed.), *Archaeology as political action* (pp. 188–221). Berkeley: University of California Press.

McMillon, B. (1991). *The archaeology handbook: A field manual and resource guide.* New York: Wiley.

Murdock, B. (1992). Lime disease prevention. *SOPA Newsletter, 16*(7), 1–3.

Neumann, T. W., & Sanford, R. M. (2001). *Practicing archaeology: A training manual for cultural resources archaeology.* Walnut Creek: AltaMira Press.

Nisser, M. (1983). Industrial archaeology in the Nordic countries, viewed from Sweden. *World Archaeology, 15*, 137–147.

Odegaard, N., & Sadongei, A. (Eds.). (2005). *Old poisons, new problems: A museum resource for managing contaminated cultural materials.* Walnut Creek: Alta Mira Press.

Oviedo, B. (2005). El patrimonio industrial en México. 20 años de estudio, rescate, reutilización y difusión. *Industrial Patrimony, 13*(7), 25–40.

Palmer, M. (2000). Archaeology or heritage management: The conflict of objectives in the training of industrial archaeologists. *IA: The Journal of the Society for Industrial Archaeology, 26*(2), 49–54.

Palmer, M. (2010). Industrial archaeology and the archaeological community: Fifty years on. *Industrial Archaeology Review, 31*(1), 5–20.

Palmer, M., & Neaverson, P. (1998). *Industrial archaeology: Principles and practice.* London: Routledge.

Poirier, D. A., & Feder, K. L. (Eds.). (2001). *Dangerous places: Health safety, and archaeology.* Westport: Bergin & Garvey.

Purdy, B. A. (1996). *How to do archaeology the right way.* Gainsville: University of Florida Press.

Quivik, F. (2000). Landscapes as industrial artifacts: Lessons from environmental history. *IA: The journal of the Society for Industrial Archaeology, 26*(2), 55–64.

Quivik, F. (2007). The historical significance of tailings and slag: Industrial waste as cultural resource. *IA: The journal of the Society for Industrial Archaeology, 33*(2), 35–52.

Reno, R. L. (1996). *Fuel for the frontier: Industrial archaeology of charcoal production in the Eureka Mining District, Nevada, 1869–1891.* PhD dissertation, Department of Anthropology. University of Nevada, Reno.

Reno, R. L., Bloyd, S. R., & Hardesty, D. L. (2001). Chemical soup: Archaeological hazards at Western ore-processing sites. In D. A. Poirier & K. L. Feder (Eds.), *Dangerous places: Health safety, and archaeology* (pp. 205–220). Westport: Bergin & Garvey.

Roberts, M. (2001). Beneath city streets: Brief observations on the urban landscape. In D. A. Poirier & K. L. Feder (Eds.), *Dangerous places: Health safety, and archaeology* (pp. 157–168). Westport: Bergin & Garvey.

Robertson, D. (2006). *Hard as the rock itself: Place and identity in the American mining town.* Boulder: University Press of Colorado.

Roskams, S. (2001). *Excavation.* Cambridge manuals in archaeology. Cambridge: Cambridge University Press.

Saunders, C., & Chandler, S. R. (2001). Get the lead out. In D. A. Poirier & K. L. Feder (Eds.), *Dangerous places: Health safety, and archaeology* (pp. 189–204). Westport: Bergin & Garvey.

Scarlett, T. J. (2007). Teaching the African diaspora and its consequences: Thoughts about entangling education. *African Diaspora Archaeology Newsletter.* December 2007. Digital Resource. African Diaspora Research Network. Retrieved May 25, 2011, from http://www.diaspora.uiuc.edu/news1207/news1207.html#4

Seely, B., & Martin, P. (2006). A doctoral program in industrial heritage and archaeology at Michigan Tech. *CRM: The Journal of Heritage Stewardship, 3*(1), 24–35.

Shackel, P. A. (1996). *Culture change and the new technology: An archaeology of the early American industrial era.* New York: Plenum Press.

Smith, C., & Burke, H. (2007). *Digging it up downunder: A practical guide to doing archaeology in Australia.* New York: Springer.

Speight, S. (1998). Digging for history: Archaeological fieldwork and the adult student, 1943–1975. *Studies in the Education of Adults, 30*(2), 68–85.

Speight, S. (2004). Teachers of adult education in British universities, 1948–1998. *Studies in the Education of Adults, 36*(1), 111–127.

Storm, A. (2008). *Hope and rust: Reinterpreting the industrial place in the late 20th century.* Stockholm Papers in the History and Philosophy of Technology, Royal Institute of Technology, Stockholm.

Symonds, J. (2004). Historical archaeology and the recent urban past. *International Journal of Heritage Studies, 10*(1), 33–48.

Symonds, J. (2006). Tales from the city: Brownfield archaeology, a worthwhile challenge? In A. Gren & R. Leech (Eds.), *Cities in the world, 1500–1700* (pp. 235–248). Leeds: Maney Publishing.

Thomas, D. H. (2000). *Skull wars: Kenewick man, archaeology and the battle for native american identity.* New York: Basic Books.

Tringham, R. (1991). Households with faces: The challenge of gender in prehistoric architectural remains. In J. M. Gero & M. W. Conkey (Eds.), *Engendering archaeology: Women in prehistory* (pp. 93–131). Oxford: Blackwell Publishers.

Walker, M., & Saitta, D. J. (2002). Teaching the craft of archaeology: Theory, practice, and the field school. *International Journal of Historical Archaeology, 6*(3), 199–207.

Walsh-Haney, H., Freas, L., & Warren, M. (2008). The working forensic anthropology laboratory. In M. W. Warren, H. A. Walsh-Haney, & L. E. Freas (Eds.), *The forensic anthropology laboratory* (pp. 195–212). Boca Raton: CRC Press.

Weisberger, J. (2003). Industrial archaeology masters program, Michigan technological university: Leading the way in a developing genre. *Journal of Higher Education Strategists, 2,* 201–206.

White, P. J. (2006). Troubled waters: Timbisha shoshone, miners, and dispossession at warm spring. *IA: The Journal of the Society for Industrial Archaeology, 32*(1), 4–24.

Zimmerman, L. J., Singleton, C., & Welch, J. (2010). Activism and creating a transnational archaeology of homelessness. *World Archaeology, 42*(3), 443–454.

Part III
Underwater

Chapter 9
The University of West Florida's Maritime Field School Experience

John R. Bratten

9.1 Introduction

Maritime Archaeology at the University of West Florida (UWF) is dedicated to the preservation of submerged cultural resources in Northwest Florida and the Southeast United States. However, it is only with the combined effort of our field school students that this is possible. Students participating in UWF maritime field methods course have experienced strong currents, zero-visibility waters, jellyfish, and an occasional sighting of alligators. In 2010, they adapted to effects from the Deepwater Horizon oil spill. However, UWF students have also experienced a strong introduction to underwater excavation, ship recording, and remote sensing techniques.

In conjunction with the city of Gulf Breeze, Florida and the Florida Division of Historical Resources (FDHR), UWF conducted its first underwater shipwreck excavation near Deadman's Island in 1989 (Bense 1988). The excavation of the Deadman's Shipwreck, an abandoned British sloop from the British period in Western Florida (1763–1781), provided valuable scientific information and served as a resource to educate the public about the region's unique underwater resources. The project was also the catalyst to involve UWF anthropology students in maritime archaeology. Since that time, UWF has offered fifteen additional maritime field schools. The length of the experience has ranged from 6 to 13 weeks and the number of students participating has varied from 5 to 35. UWF has developed several methods to accommodate large numbers of students and provide them with a safe, educational, and rewarding field school experience.

J.R. Bratten (✉)
Department of Anthropology, University of West Florida, Pensacola, FL 32571, USA
e-mail: jbratten@uwf.edu

H. Mytum (ed.), *Global Perspectives on Archaeological Field Schools:* 147
Constructions of Knowledge and Experience, DOI 10.1007/978-1-4614-0433-0_9,
© Springer Science+Business Media, LLC 2012

9.2   The University of West Florida Field Methods Courses

Since the 1980s, UWF has offered a terrestrial field methods course at both the graduate and undergraduate levels. Compared to many university field schools, UWF field courses are often of long duration. For many years, the courses were offered for the entire summer session, lasting for 13 weeks. Recently, this period has been shortened to 10 weeks so that students and instructors can spend some time with their families before the start of the fall semester. UWF's archaeology program was created by Dr. Judith Bense, who considered that shorter field schools only allowed students to be introduced to methods; a longer field school would give students the time to not only learn methods, but also apply them. In this way, it was hoped that graduates would leave UWF with the necessary skills to begin work with cultural research management agencies.

Both prehistoric and historic terrestrial sites are excavated and often students have a choice in the age or period of the excavation or survey site. Undergraduate anthropology students enroll for 9 h of credit and graduate students enroll for 3 h of credit. Graduate students complete the course at the supervisory level and are provided with advanced training in survey, testing, and site excavation. They are also trained in project planning, budgeting, supervision, and integration of information recovered from the field. For the past 2 years, a biological archaeology field school has also been offered.

9.2.1   Maritime Field Methods

A maritime field methods course similar to the terrestrial field schools provided by UWF is offered to both undergraduates and graduate students during the summer. All graduate students enrolling in an advanced field methods course must have had an acceptable field methods course at the undergraduate level, however. For this reason, many students who apply to our graduate programs and are interested in the maritime field methods course are often required to complete the undergraduate maritime field school as a prerequisite. Usually, they enroll in the undergraduate course their first summer and enroll during the graduate version in their second summer in residency. Entry to the course is by application which is used to determine that prospective students have proper diving certification (PADI, NAUI, SSI, YMCA, etc.) and have had a prerequisite upper-level archaeology course such as UWF's Principles of Archaeology (ANT 3101). Students from within and outside the university are encouraged to apply.

The undergraduate maritime field methods course (ANT 4835) is designed to provide a structured hands on experience including training in both field and laboratory methods. Methods taught include site control grids, setting up excavation units, basic excavation techniques, use of hand tools, identification of ship structure and features, screening techniques, field documentation, principles and use of field instruments, and field conservation procedures. The graduate version of the course,

Advanced Archaeological Field Methods (ANG 6824) course, is similar to the undergraduate version, but like its terrestrial counterpart includes additional training in project planning, budgeting, supervision, and integration of information recovered from the field. Both maritime courses are taught concurrently, with graduate students assigned to supervise the undergraduate students. The ratio of graduate to undergraduate students varies by year, but the number of undergraduate students usually predominates by a ratio of 2:1 or 3:1.

Subject to funding, graduate students are paid at least one-quarter time (10 h/week) to compensate them partially for the long hours that they are required to work in the overall effort. In addition to supervising undergraduate students in actual fieldwork, graduate students are expected to ensure that any boats and trailers needed for the day's activities are in good working order, gasoline is on hand, and that all equipment needed for the day's fieldwork is loaded into the boats or transport vans. With a large field school, this preparation may entail considerable work if, as often the case, three boats and trailers are required along with dredge pumps and their associated hoses and fittings, scuba tanks, and safety equipment (first aid kits, oxygen administration kits, back boards, life jackets, etc.). In addition to equipment, various form boxes, archaeological recording equipment (flexible tapes, folding rules, cameras, mylar, pencils, etc.), personal diving gear, water, and lunches must also be carried. This process is usually slow and unwieldy for the first days of field school, but with the use of morning and afternoon briefings, assignments are quickly distributed between the graduate and undergraduate students. The process soon improves, and with the use of checklists, becomes efficient.

In recent years, one or two senior graduate students who have already completed their advanced field methods course have been hired to act as Field Directors in Training. These students are paid full-time (40 h/week) and given a number of responsibilities including direct supervision of the graduate students. As old hands in the process, they provide wisdom and experience for the entire operation. By this provision of staff, everyone becomes informed and effective very quickly so that fieldwork can begin as early as possible. This is also helped by requiring the graduate supervisors and field directors to begin the planning and equipment organization several weeks before the actual start of field school. Indirect and direct supervision is also provided by the course instructors, which has consisted of two faculty members for all but three of the years of field schools.

9.2.2 Combined Field Methods

At the request of a student who wanted to gain field method experience in both areas of terrestrial and maritime excavation, UWF began to offer a combined field methods course (ANT 4121) in late 1990s. The course consists of 5–6 weeks of on-site training on a terrestrial site followed by 5–6 weeks training on an underwater site (usually a shipwreck excavation and/or remote sensing project). Initially, this course was an option and only a few students choose to enroll for several years. Gradually, the course increased in popularity and enrollment numbers increased. Enrollments in the regular maritime field methods course also increased and for several reasons. In 1998, George F. Bass published an article in *Archaeology* magazine entitled: "History Beneath the Sea: The Birth of Nautical Archaeology." On page 52, Bass (1998) states that:

> Given that we now have access to what one scholar termed the world's greatest museum, why is it that, as the twentieth century draws to a close, only three universities in the United States, Texas A&M University, with its affiliated Institute of Nautical Archaeology, the University of West Florida, and East Carolina University, train marine archaeologists? And why are they the only universities that conduct shipwreck surveys and excavations—and only the former on a global scale?

It is without question that this mention in *Archaeology* drew awareness to UWF and added students to our programs. At the same time, UWF maritime field schools had made national news with the announcement that one of its sites, the Emanuel Point shipwreck, was formally associated with the 1559 colonization attempt of Florida by Don Tristán de Luna y Arellano. Potential students learned that they would be allowed to participate in the excavation of significant shipwrecks if they enrolled in a UWF field school.

In 2003, UWF offered two new programs through the Anthropology Department, an MA in traditional Anthropology (prior to 2003 UWF offered only an MA in Historical Archaeology), and an interdisciplinary BA program in Maritime Studies. Enrollment in the both these programs generated additional students who either wanted or were required to complete a field school experience. As a result of this increased demand for maritime field methods courses, the combined field school has been offered as the only choice for undergraduate students and graduate students needing the prerequisite for the past several years. So that as many students can have the underwater excavation experience, one half of the students participate in the first 5-week session as maritime students and then switch to either of the two terrestrial field schools, with others the two elements in the reverse order. Initially, some students who were extremely desirous of underwater excavation experience were not completely satisfied with this only option, but changed their minds to a favorable opinion of the combined course upon completion and also learned valuable terrestrial excavation skills. With this method, more than thirty students were trained in maritime field methods in 2009 and 2010.

9.3   Dive Safety Program

The UWF Dive Safety Program began with the creation of a guide for scientific diving in 1993. It was adapted from the U.S. Navy Diving Manual, the National Oceanic and Atmospheric Administration Diving Guide, and the American Academy of Underwater Sciences guide to scientific diving and has been revised and updated as needed. UWF's dive safety program seeks to "ensure that all scientific diving under the auspices of UWF and [its allied] Marine Services Center (MSC) is conducted in a manner that will maximize the protection of scientific divers from accidental injury and/or illness" (Marine Services Center 2010). UWF's Dive Safety Program is directed by the Dive Safety Officer (DSO) whose job is to develop and implement the Scientific Diving Program. The DSO is the primary support person for diving from our facility and is responsible for the operation and maintenance of the dive equipment and its locker. Working with principal investigators, he also directs and instructs staff, students, and research divers in the safe and efficient operation of marine diving and often accompanies student divers in the early stages of the maritime field schools. The DSO verifies and maintains diver certification and project dive logs and is also responsible for enforcing diving safety regulations and approving all dive plans.

The DSO reports to UWF's Marine Services Executive Committee, an administrative group consisting of members from each UWF department engaging in underwater activities and a financial officer. Meetings are held bimonthly. The DSO also reports to UWF's Diving Control Board which includes members from active departments and UWF's Associate Director for Recreation and Sports Services. The Diving Control Board meets annually to review diving activities. Their responsibilities include:

1. Acting as a board of appeal to consider diver-related problems.
2. Recommending changes in policy and amendments to the UWF Scientific Diving Guide.
3. Establishing and/or approving student and instructor training programs.
4. Recommending new equipment or techniques.
5. Recommending, establishing, and/or approving facilities for the inspection and maintenance of diving and associated equipment.
6. Reviewing the DSO's performance and program.
7. Sitting as a board of investigation to inquire into the nature and cause of diving accidents or violations of the UWF Scientific Diving Guide.

9.4 Boats, Facilities, and Platforms

The UWFs MSC serves all the departments, centers, and institutes involved in university-sponsored projects concerning underwater research and provides diving platforms, research vessels, and support staff. MSC possesses a number of vessels and platforms which are suitable for a variety of uses. These include pontoon boats and a research barge for coastal and estuarine research (Fig. 9.1), as well as several craft which are suited for operations in deeper Gulf of Mexico waters. To offset the cost of maintenance and repair, each project pays a daily usage fee for any of the vessels. The cost, paid out of research grant funds if available, varies from $45 (research barge) to $400/day (offshore vessel). This cost can be considerable for a 10–13-week underwater field school when multiple vessels are required to carry a large number of students to and from the sites and operate an additional vessel

Fig. 9.1 The UWF research barge

involved in remote sensing operations. The expense is well worthwhile, however, as it ensures that properly maintained vessels are always available, including all their necessary safety equipment.

MSC is supervised by a full-time person whose salary and benefits are paid out of the university's sponsored research program. For a number of years, MSC has been headed by Steve McClin, an individual with considerable experience in marine repair and metal fabrication. His skills have been extremely useful to our field schools. This is especially true in the modifications he has made to our research barge. The barge serves as our project diving platform and offers plenty of shaded space for a large number of students and their associated gear, both diving and excavation. It would be extremely difficult for UWF to host a large number of students without the availability of a research barge that can be moved from site to site. Fortunately, most of our maritime field courses focus on shipwrecks or abandoned vessels in the relatively shallow waters of Pensacola Bay, Pensacola Sound, or nearby rivers. As such, it is easy to move the barge to a new site each season and anchor it alongside a site with a three-point mooring system. In the event of an approaching hurricane or tropical storm, the barge can moved to a safe location and later repositioned back on site.

9.5   Scientific Diver Training Week

All maritime field school courses and the combined course are preceded by a week of "Scientific Diver Training." This training is designed to assess the student's diving skills, provide training in first aid and cardiopulmonary resuscitation (CPR), and practice dry land training exercises supplemented by lectures. A typical timetable of activities for a Scientific Diver Training week is laid out in Table 9.1.

Table 9.1   A typical timetable of activities for a Scientific Diver training week

Day 1
   8:00am – 11:00am: Introduction
      Introduction to the diving program (facilities, personnel, rules, etc.)
      Review diving paperwork (waivers, etc.)
      Introduction to the University of West Florida's Scientific Diving Guide, brief/debrief, hand
         signals and communication, emergency procedures protocol, responsibilities of a safety
         diver, etc.
      Tour dive locker
      Checkout gear to students
   11:00–12:00: Lunch
   1:00– 4:00pm: Dry Land Archaeology Training
      Low and high tech survey methods
      Knots
      Circle search/metal detecting techniques
      Excavation dredge setup and maintenance
      Compass navigation
      Mapping and measuring
      Field artifact collecting and tagging
Day 2
   8:00–11:00am: Background Information and Site Presentations
      Introduction to ship construction and recording features
      Lay baselines, conduct offset measurements on small boats
      Artifact identification, dredge spoil sorting
      Field books
      Piece plotting artifacts
      Background (history/archaeology) of sites chosen for field school
      Tour of associated exhibits (e.g. Emanuel Point Shipwrecks)
   12:00–4:00: Free Time
      Let students get any last minute gear
      Extra time with helping students with archaeology skills
Day 3
   8:00am – 12:00pm: Safety Training
      Cardiopulmonary resuscitation
      First aid
   12:00–1:00pm: Lunch
   1:00pm – 3:00: Confined Water, SCUBA Checkout
      Swim test 230 m (any style, no time, nonstop)
      Tread water (5 min legs and arms, 1 min head/hands out of water)
      23 m underwater swim (1 breath w/kickoff from side of pool)
      Snorkel checkout
      500 m surface swim with snorkel gear (wet suits are permitted)

(continued)

Table 9.1 (continued)

SCUBA Checkout
  Group 1: Tired diver/ unconscious diver tow and push, cramp relief, ditch and don gear on
    surface
  Group 2: Entries, mask clears/recovery, regulator clear and recovery, ditch and don Scuba
    Gear Underwater
  Group 3: Air sharing skills, controlled emergency swimming ascents, buoyancy control
  3:00pm–5:00pm: Confined Archaeology Training
    Knots
    Circle search/metal detecting
    Dredge setup and breakdown
    Compass navigation
    Mapping and measuring
Day 4
Location: Marine Services Center (bring bag lunch)
  8:00am–9:30am: Load boats and gear
    Travel to UWF barge for open water checkout and training session
  9:30am–11:30am: Open water SCUBA checkout
Snorkel Checkout
  500 m with snorkel gear
SCUBA Checkout
  Group 1: Tired diver/ unconscious diver tow and push, cramp relief, ditch and don gear on
    surface
  Group 2: Entries, mask clears/recovery, regulator clear and recovery, ditch and don SCUBA
    gear underwater
  Group 3: Air sharing skills, controlled emergency swimming ascents, buoyancy control
  11:30am–2:30pm: Open Water Archaeology Training
    Knots
    Circle search/metal detecting
    Dredge setup and breakdown
    Compass navigation
    Mapping and measuring
  2:30pm– 3:30pm
    Return to Marine Services Center: Clean gear and hang to dry
Day 5
Reserved for students who need additional help with skills

All UWF Scientific Divers must sign a form indicating that they will adhere to the regulations, requirements, and procedures outlined in UWF's Scientific Diving Guide. This includes, but is not limited to, diving within certification limits; reporting unsafe practices, injuries, or incidents to the DSO; following dive plans; maintaining proper buoyancy; maintaining personal dive gear; and terminating all dives with enough air in tank to surface with at least 34 BAR (500 PSI).

At the completion of the Scientific Diver training, the student must satisfy the DSO of his/her ability to perform the following:

1. Demonstrate proficiency in buddy breathing as both donor and receiver, or share air situation.
2. Enter and leave open water or surf, or leave and board a diving vessel, while wearing SCUBA gear.

3. Demonstrate, where appropriate, the ability to maneuver efficiently in the environment, at and below the surface.
4. Complete a simulated emergency swimming ascent.
5. Demonstrate clearing of mask and regulator while submerged.
6. Demonstrate ability to remove and replace equipment while submerged.
7. Demonstrate ability to achieve and maintain neutral buoyancy while submerged.
8. Demonstrate techniques of self-rescue and buddy rescue.
9. Navigate underwater.

All UWF Scientific Divers must submit a statement from a licensed physician, based on an approved medical examination, attesting to the applicant's fitness for diving. For students under the age of 30, this form is valid for a period of 3 years. For those 30 or over, the form is valid for 2 years. After the 1-week scientific diver training, students are in the field every day from 7 AM until 3 PM. When not diving (Fig. 9.2), students perform topside duties such as dive tending and support, artifact recording, and database entry. Students may also be called upon to participate in the conservation and laboratory analysis of recovered material.

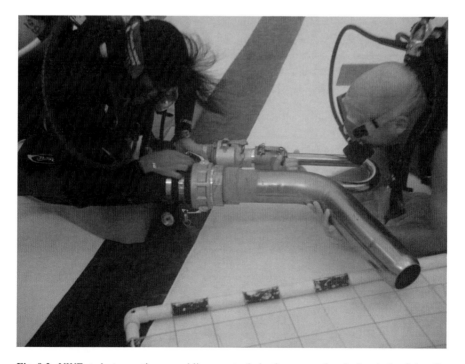

Fig. 9.2 UWF students practice assembling a water induction excavation dredge during Scientific Diver training week

9.6   Grade Determination

Both objective and subjective factors determine the student's grade in the maritime field courses. The objective factors are: acquisition of field skills; accuracy, completeness, and relevance of field notes (graded weekly by supervisors); accuracy, completeness of field forms, and drawings (reviewed by supervisors); and attendance. The subjective factors are: team cooperation, leadership, reliability, responsibility, improvement of skills/initiatives, language, professional demeanor, and interaction with the public. The weight of factors used for grade determination is field skills and conduct (40%), documentation (40%), and intangibles (subjective factors, 20%).

9.7   The UWF Maritime Archaeological Field Methods Courses

UWF offered its first terrestrial field school in 1986 and its first underwater field school in 1989. The maritime experience has varied from remote sensing surveys to phase III excavation. Two types of maritime field methods are taught: site survey and basic excavation skills. These skills include using vertical and horizontal grids and datums, setting up excavation units, basic excavation techniques, use of hand tools, identification of ship structure and features, screening techniques, field documentation, principles and use of field instruments, and field conservation procedures. Other activities include instruction in the use of remote sensing survey equipment such as magnetometer, side-scan-sonar, and subbottom profiler, as well as specific techniques for hull recording, underwater photography, site assessments and dives on known historic wrecks located in the area. Fieldwork is supplemented by lectures and discussions on themes ranging from the colonization of Northwest Florida, maritime landscapes, and economic maritime connections in the Gulf region. The range of sites and field conditions experienced on UWF field schools can be demonstrated by considering the locations of the field school over the last two decades. This review also demonstrates the types of site that is feasible for a field school and the educational, research, and management outputs that can be derived from this form of fieldwork.

9.7.1   The Deadman's Island Shipwreck

Following the report of a shipwreck just off Deadman's Island near Gulf Breeze, Florida, preliminary investigations of the site occurred in 1988 under the direction of Florida's state underwater archaeologist, Dr. Roger C. Smith. Based on the energetic participation of two UWF students, UWF decided that a maritime field school would be offered in 1989. All field and laboratory work would be performed by students interested in marine archaeology. The field school lasted for 6 weeks and Smith utilized the participation of twelve students to remove the overlying sand from ship's timbers, recover and conserve artifacts, and produce a site plan and report (Smith 1990). The field school was considered a success and planted the seed for future underwater investigations by the university. The student work was brought to the attention of the public when artifacts from the site were placed on public display in Gulf Breeze and a local newspaper featured the work in an article. Following the news story, there was a public outcry when it was discovered that weekend vandals had removed a number of timbers from the site and left them to dry out in the sun. The incident made it apparent that the public valued their historical resources and were pleased to see student involvement in their documentation and preservation. Since the original public outcry in 1989 against vandalism of archaeological sites, UWF has been fortunate that none of its other projects have been subject to any destructive activities or site disturbance.

Archaeology at UWF has always been conducted in the public eye. From the earliest excavation projects, volunteers from the community have always been welcome and have helped with the overall outcome in many ways. Both diving and nondiving volunteers have also participated in the underwater field schools, working alongside the students. Volunteers adhere to the same diving rules and participate in Scientific Diver training. Students keep the public at large informed of their work through a weekly Internet blog and by presenting talks to local groups such the Pensacola Archaeological Society or other community organizations.

9.7.2   Emanuel Point I Shipwreck

Pensacola's rich underwater resources were placed in the national spotlight with the discovery of the Emanuel Point shipwreck in 1992 by the Florida Bureau of Archaeological Research. The state team, led again by Smith, located the wooden remains of the shipwreck during what was proposed to be a statewide survey of Florida's shipwrecks, county by county, with Escambia County in extreme Northwest Florida chosen as the starting point. Because of the vessel's assumed significance and its potential association with the 1559 Spanish colonization attempt of Florida, Smith prepared a 5-year plan which included the gradual development of a program in marine archaeology at UWF (Smith et al. 1995:xii) and future UWF field schools. UWF agreed to become an academic partner in the multiyear excavation of the Emanuel Point ship (Smith et al. 1995:xiii). UWF's second maritime underwater field school began in May 1993 with eleven students (graduate and undergraduate) from several universities.

In 1997, UWF offered its third underwater field school and the first with its own instructors. Five students enrolled and participated in the second excavation campaign on the Emanuel Point Ship. Aided by the work of these students, UWF and state archaeologists firmly associated the Emanuel Point shipwreck with the fleet of Don Tristán de Luna y Arellano (Smith et al. 1998). Luna's fleet had been struck by a hurricane in 1559 during the first major European attempt to colonize what is now modern day Florida. Students and professional archaeologists recovered more than 3,500 artifacts from the vessel during the two campaigns of fieldwork. Portions of the ship's bow, stern, and midship areas were opened up to reveal the structure of a large wooden vessel, likely one of the fleet's large galleons. A variety of plant and animal remains (including rats, mice, and cockroaches) were collected along with stone and iron cannonballs, copper cooking cauldrons, and a single coin minted between the years 1471 and 1474. Other finds included the discovery of a steel breast plate, Aztec ceramics, and a small wooden silhouette of a ship carved in the shape of a galleon. From 1997 onward, enrollment in the maritime field methods steadily increased. Student analysis of the shipwreck site also resulted in the production of five MA theses (Scott-Ireton 1998; Fossum 2001; Rodgers 2003; Collis 2008; Lawrence 2010).

9.7.3   Santa Rosa Island Shipwreck and Catharine

Following the field schools at the Emanuel Point shipwreck, UWF has offered an underwater field methods course every summer to date. In 1998, a 6-week field school allowed students to participate in the archaeological recording of the *Catharine* and the Santa Rosa Island shipwreck. The former was a Norwegian lumber ship that sank in 1894 during a winter storm in the Gulf of Mexico. The latter was a large vessel located in Pensacola Sound in 3.5–5.5 m of water.

At the close of the 1998 investigations, UWF students and researchers established a tentative date for the Santa Rosa Island Wreck. Pottery types fell within a 1680–1720 date range. Preliminary analysis of the hull remains indicated that the vessel exhibited characteristics similar to those of previously investigated eighteenth-century ships. Wood samples taken from a variety of hull members revealed that the ship was constructed exclusively of New World hardwoods--specifically, Spanish cedar and mahogany (*Swietenia* sp.). The massive size of individual timbers in the ship's hull indicated that the wreck was once a large, oceangoing vessel, perhaps engaged in commerce or defense of European interests in the colonial New World.

Although the early field investigations revealed much about the shipwreck site, many important questions pertaining to the vessel's function, identity, nationality, and history remained to be answered. Consequently, UWF hosted a second field school at the Santa Rosa Island Wreck during the summer of 1999, with the goal of locating one end of the ship and gathering enough evidence to aid in the identification of the vessel. Over the course of a 13-week field school, faculty, staff, and students located, excavated, and recorded the bow of the shipwreck. After removing nearly 1.5 m of overlying sand, field school students and volunteers plotted and recovered hundreds of artifacts of many different varieties (Fig. 9.3).

Fig. 9.3 UWF students examine the mahogany keelson and mast step of the Santa Rosa Island shipwreck in Pensacola Sound, Florida

Additional funding from the State of Florida allowed fieldwork to continue with 13-week field schools in 2001 and 2002. During this period, students excavated and recorded the starboard side of the Santa Rosa Island vessel and placed several trenches in key portions to determine the amount of hull preservation. Project members recovered more than 2,300 artifacts from the wreck during fieldwork.

UWF Historical Archaeology graduate student James W. Hunter III initiated historical research into the vessel's identity. James' research and the archaeological interpretation reveal that the site most likely represents the remains of the *Nuestra Señora del Rosario y Santiago Apostol*, a large frigate and former flagship of the Spanish Windward Fleet, which had patrolled Gulf and Caribbean waters. *Rosario* was lost in a 1705 hurricane shortly after arriving at Presidio Santa María de Galve (1698–1719), near the modern city of Pensacola, Florida (Hunter 2001).

In terms of the field school experience, the Santa Rosa Island Shipwreck project proved to be one of the more challenging to students in terms of work environment. Due to its location, the wreck is subject to strong currents depending on the particular moon phase and time of day. On a few days, early in the field school, some students struggled against a strong current and found it necessary to grab on to a floating safety rope placed for that very purpose so that they could be pulled back to the excavation platform. Very quickly, most students learned to adapt to the current, use down lines, and plan their exit from the site without being swept away to the safety line. Students also learned that once they reached the bottom, the effect of the current was mostly above them and would not affect their excavation work to any great extent once they were in their units. Work at the site of the *Catharine* allowed students to experience working conditions in the Gulf of Mexico and provided information for an MA thesis topic (Burns 2000).

9.7.4   Hamilton's Shipwreck

In the year 2000, students enrolled in Maritime Field Methods concentrated their efforts on the recording of a snapper schooner named after the Pensacola crab fisherman who discovered the wreck, Frank Hamilton. Hamilton's Wreck is located in only 1.3 m of water in Pensacola Bay near the Naval Air Station and dates to the early twentieth century. Students were introduced to ship recording using the Direct Survey Method (DSM). Developed by Nick Rule, DSM is the use of direct tape measurements from datum points of known three-dimensional coordinates, redundant data to identify and qualify errors, and the use of computer programs to process the data and find best fit solutions for the points being surveyed (Rule 1989). Due to the shallow depth of the site, students also learned to work with surface supplied air using a hookah system. Although freed from carrying a scuba tank, the hookah system required time for adaptation as it necessitates working with an attached hose and air intake which requires a little more effort on the part of the student. The project was documented by participating student Moore (2002) in his MA thesis.

9.7.5   Snapper Wreck and Rhoda

In 2003, UWF field school students documented the remains of an abandoned vessel known as the "Snapper Wreck" in the Blackwater River, near Bagdad, Florida. Although the specific name of the shipwreck remains a mystery, it was determined that the vessel was probably a two-masted schooner, locally known as a snapper smack (Raupp 2004). UWF also examined a shipwreck associated with Pensacola's lumber industry. The Rhoda sank in a violent storm in 1882 and involved student recording over a large area of Pensacola Sound (Rawls 2004). Both ships provided students with a glimpse of the lumber and fishing vessels in use during Pensacola's commercial period, and the lives of the men that lived and worked aboard. Work at the Snapper Wreck presented a new challenge to some students, diving in zero to low-visibility river water.

9.7.6   The Shields Point Vessels

During the summers of 2004, UWF students recorded the sunken remains of seven vessels in Blackwater Bay. These included four schooner barges, a small steam-powered tugboat, and two other sailing type vessels. Most of these vessels have been firmly connected to Pensacola's historic lumber industry and were abandoned sometime in the late 1920s or early 1930s following the decline of the industry. Four of these vessels were identified by name, *Dinty Moore*, *Geo. T. Lock*, *Guanacaste*, and the locally-built *Palafox*. The other two sailing vessels probably represent the remains of fishing smacks. As with the earlier field schools, students had to face the challenge of working in low-visibility areas due to the high concentration of tannic acid in the water and a visit from a relatively harmless alligator. Relying on the extensive use of field student-produced DSM data, three additional MA theses were produced (Holland 2006; Pickett 2008; Sjordal 2007).

9.7.7 Deadman's Island and the Seminole Wreck

In 2005 and a portion of the 2006 maritime field school, UWF field school students returned to Deadman's Island to concentrate on a remote sensing survey with magnetometer and side-scan-sonar to record structures associated with the island's history as a quarantine station and careening station. Students participated in near shore recording projects such as the remnants of a marine railway, ballast piles, pilings, timbers, a boiler, crates, and ceramic finds and several abandoned vessels or shipwrecks. Students learned circle search techniques, and offset and baseline recording methods. In addition to learning how to deploy and use the remote sensing equipment, students learned to HYPACK, a hydrographic survey software program created expressly for survey applications, to plan and process both magnetometer and side-scan-sonar survey data. Student-collected data were compiled and incorporated into a MA thesis analysis, the maritime landscape of Deadman's Island (Jordan-Greene 2008).

In 2006, students were also given the opportunity to record features of a virtually intact side-wheel steamship located in the Blackwater River near Seminole, Alabama. Challenges to the site were low-visibility water at the vessel's lower depths and the preponderance of recordable features including the ship's deck, boiler, engine, and almost all it paddlewheel assemblage, all of which will be documented in a forthcoming MA thesis (Abrahamson 2011).

9.7.8 The Emanuel Point II Shipwreck

UWF archaeologists and field school students conducted a systematic magnetometer survey near the site of the Emanuel Point shipwreck in 2006. As a result, two previously undocumented shipwrecks were discovered. Initially designated as magnetometer targets, Target 2 was determined to be the remains of a shipwreck carrying a large quantity of bricks and dated to a period slightly before the Civil War. Preliminary test excavation on a second target, Target 17, revealed an extensive stone ballast pile covering well-preserved wooden hull remains. Recovered artifacts including Spanish ceramics and strips of lead hull sheathing suggested that the vessel dated to the Pensacola's first Spanish period (1559–1719) and might also be associated with Luna's colonization attempt in 1559. The vessel is now referred to as the Emanuel Point II shipwreck (EP II).

During the summers of 2007–2010, UWF conducted further excavation on EP II in the form of 10–12-week field schools. This work defined the site's extent and confirmed the vessel's nationality and historical associations with Luna and Pensacola's first European settlers. Discovery of this second sixteenth-century Spanish ship, approximately 400 m west of the first Emanuel Point site, has led to an unprecedented comparative study of two vessels from an early colonization fleet. The excavations have provided new information concerning sixteenth-century ship-building practices and identify who Luna may have brought with him to Florida.

Unfortunately, work at EP II was halted midway during the 2010 maritime field school. Not long after the news of the Deepwater Horizon explosion and oil spill, questions about the possibility of oil and/or oil dispersants entering Pensacola Bay and affecting the site were raised. By the beginning of our June 1 course, oil sheen was reported less than 10 miles off Pensacola Beach. Prevailing winds forecasted the arrival of outer edge of the slick, and emergency crews began to place kilometers of boom in an effort to prevent the oil from entering Pensacola's inland waterways. Pensacola Beach was closed on June 24 and a health advisory was issued for the county in which we were conducting our field work. On the same day, oil sheen arrived at the site of the EP II shipwreck and all diving operations in the bay were suspended. It was during this week the students enrolled in combined maritime field school made their switch. Unfortunately, for these students, the oil spill curtailed their work on the EP II wreck. The remainder of the field school was spent in the oil free waters of the Blackwater River where the students documented the wrecks mentioned above, learned DSM techniques, documented two additional vessels, and discovered a new wreck with the remote sensing equipment. At the time of writing, Pensacola Bay has been declared free of any oil or dispersants that would affect a summer 2011 field school on the EPII site. A large student enrollment will aid in the excavation of the vessel's midships area.

9.8 Conclusion

In addition to the wealth of information that our local shipwrecks are providing to archaeologists, historians, and the interested public, the projects have served as an effective field laboratory for our students and those who have enrolled from a dozen colleges and universities in a variety of academic disciplines including marine archaeology, chemistry biology, and history. Nearly two hundred students and volunteers have worked alongside professional archaeologists studying the shipwrecks since 1989. Students and volunteers have also played an active role in our conservation laboratory by documenting artifacts and compiling databases. They have also put in countless hours removing salts and applying resins and other preserving agents to protect fragile artifacts. Still others are contributing to the project's historical research and ship analyses with the production of conference presentations, class reports, and masters theses related to the studies of the vessels' food remains (plant and animal), ceramic assemblage, and hull analyses. The UWF's Anthropology and Maritime Studies programs have grown and benefited as a direct result of the maritime field schools. Continued fieldwork, conservation, and survey efforts will prove equally beneficial for the field, the university, its students, and future volunteers. Pensacola is known as the "City of Five Flags" and, as such, its waters are rich with materials from this diverse cultural heritage and the public welcomes and appreciates the work of our students and faculty.

Acknowledgments   The author would like to thank the numerous field school students and volunteers who have participated in the UWF Maritime Field Schools. Without their help, none of our maritime research efforts would have been possible. Thanks also to my colleagues Gregory D. Cook, J. "Coz" Cozzi, and Roger C. Smith for instructional leadership and development of our maritime field schools. The many projects UWF has undertaken since 1997 would not have been successful without the assistance of the Marine Services staff, Director Steve McLin; DSOs, Fritz Sharar, Dwight Gievers, Lloyd Oubre, Mike Lavender, Jason Raupp, Paul Sjordal, and Keith Plaskett. I also want to thank Elizabeth Benchley and the UWF Archaeology Institute for supporting the maritime archaeology field schools.

References

Abrahamson, W. A. (2011). Whiskey and windowpanes: An archaeological investigation of an Eastern Coastal Side Paddle-Wheel Steamboat at Seminole, Alabama. Unpublished MA thesis, Department of Anthropology, University of West Florida, Pensacola.

Bass, G. F. (1998). History beneath the sea: The birth of nautical archaeology. *Archaeology, 51*(6), 48–53.

Bense, J. A. (1988). *Deadman's shipwreck: Preliminary investigation and evaluation.* Office of Cultural and Archaeological Research, Reports of Investigation, No. 18, University of West Florida, Pensacola, Florida.

Burns, J. M. (2000). *The life and times of a merchant sailor: History and archaeology of the Norwegian Ship Catharine.* Unpublished MA thesis, Department of History, University of West Florida, Pensacola.

Collis, J. D. (2008). *Empire's reach: A structural and historical analysis of the Emanuel Point shipwreck.* Unpublished MA thesis, Department of Anthropology, University of West Florida, Pensacola.

Fossum, A. H. (2001). *Historical rationale and museum survey in support of a Maritime Museum for Pensacola.* Unpublished MA thesis, Department of Anthropology, University of West Florida, Pensacola.

Holland, L. K. (2006). *Maritime technology in transition: Historical and archaeological investigations of the Schooner Barge Geo. T. Lock (8SR1491).* Unpublished MA thesis, Department of Anthropology, University of West Florida, Pensacola.

Hunter, J. W., III. (2001). *A broken lifeline of commerce, trade and defense on the colonial frontier: Historical archaeology of the Santa Rosa Island Wreck, an early eighteenth-century Spanish Shipwreck in Pensacola Bay, Florida.* Unpublished MA thesis, Department of History, University of West Florida, Pensacola.

Jordan-Greene, K. D. (2008). *A maritime landscape of Deadman's Island and Old Navy Cove.* Unpublished MA thesis, Department of Anthropology, University of West Florida, Pensacola.

Lawrence, C. L. (2010). *An analysis of plant remains from the Emanuel Point shipwrecks.* Unpublished MA thesis, Department of Anthropology, University of West Florida, Pensacola.

Marine Services Center. (2010). Dive safety information, Electronic. Retrieved September 27, 2010, from http://uwf.edu/marineservices/divesafety.cfm.

Moore, R. E. (2002). *Hamilton's wreck: An archaeological and historical inquiry into the regional maritime culture of Pensacola, Florida, 1900–1920.* Unpublished MA thesis, Department of History, University of West Florida, Pensacola.

Pickett, S. G. (2008). *Harbor kings: Analysis of shields point wreck #5 using a world systems perspective.* Unpublished MA thesis, Department of Anthropology, University of West Florida, Pensacola.

Raupp, J. T. (2004). *Hook, line, and sinker: Historical and archaeological investigations of the Snapper Wreck (8SR1001).* Unpublished MA thesis, Department of History, University of West Florida, Pensacola.

Rawls, J. K. (2004). *Time and tide wait for no one: The history and archaeology of the British Bark, Rhoda.* Unpublished MA thesis, Department of History, University of West Florida, Pensacola.

Rodgers, R. R. (2003). *Stale bread and moldy cheese: A historical and archaeological study of sixteenth-century foodways at sea using evidence collected from the Emanuel Point shipwreck.* Unpublished MA thesis, Department of History, University of West Florida, Pensacola.

Rule, N. (1989). The Direct Survey Method (DSM) of underwater survey, and its application underwater. *The International Journal of Nautical Archaeology and Underwater Exploration, 18*(2), 157–162.

Scott-Ireton, D. A. (1998). *An analysis of Spanish colonization fleets in the age of exploration based on the historical and archaeological investigation of the Emanuel Point shipwreck in Pensacola Bay, Florida.* Unpublished MA thesis, Department of History, University of West Florida, Pensacola.

Sjordal, P. G. (2007). *Wrecking the site type: The history and archaeology of ship abandonment at shields point.* Unpublished MA thesis, Department of Anthropology, University of West Florida, Pensacola.

Smith, R. C. (1990). Marine archaeology comes of age in Florida: Excavation of Deadman's shipwreck, a careened British Warship in Pensacola Bay. In T. L. Carrell (Ed.), *Underwater archaeology proceedings from the society for historical archaeology conference, 110-6.* Tucson: Society for Historical Archaeology.

Smith, R. C., Spirek, J., Bratten, J. R., & Scott-Ireton, D. A. (1995). *The Emanuel point ship: Archaeological investigations 1992–1995.* Bureau of Archaeological Research, Division of Historical Resources, Florida Department of State.

Smith, R. C., Bratten, J. R., Cozzi, J., & Plaskett, K. (1998). *The Emanuel point ship archaeological investigations 1997–1998.* Report of Investigations No. 68, Archeology Institute, University of West Florida, Pensacola.

Chapter 10
Freshwater Underwater Archaeology Field School, Good Practice, Good Science

Anne Corscadden Knox and Sheli O. Smith

10.1 Introduction

As working practitioners, how do we informally and formally assess that today's students are attaining good practice, and good science before awarding them degrees in Anthropology and/or Archaeology? In an unofficial survey of professionals in our field, we found a commonly held tenet: It takes 1 year to learn the trade and another to master it. If this is a commonly held understanding then our question comes full circle – how do we determine that students are acquiring "good practice" instead of "bad habits" that, although not intentionally passed on, are modeled and then practiced by less experienced professionals? How do we ensure that students acquire the appropriate best practice skill sets?

In the current collegiate environment, Methods and Theory courses abound within the classroom context but how many applied courses exist? Some professionals argue that Anthropology should add a fifth field, Applied Anthropology (Pyburn and Wilkes 1995). In truth, the field of archaeology would be hobbled without the experience of application of theory and methodology. Many institutions of higher learning require students to either attend a field school or accompany professionals on projects in the field. Of these field school opportunities, few are tightly structured and have carefully aligned learning targets to project goals. The result is that many of today's graduating archaeologists are expected to direct or lead field inquiry with little or no applied experience. This is simultaneously a disservice to students and to the profession.

This situation is not unique to the science of archaeology, but the inherent destructive nature of excavation leads to irrevocable losses. Therefore, fundamental to our ethical obligations as stewards of cultural heritage should be an equal dedication to instilling good archaeological habits that students will carry with them through their

A.C. Knox (✉) • S.O. Smith
PAST Foundation, 1929 Kenny Road, Columbus, OH 43210, USA
e-mail: acorscadden@pastfoundation.org

H. Mytum (ed.), *Global Perspectives on Archaeological Field Schools:*
Constructions of Knowledge and Experience, DOI 10.1007/978-1-4614-0433-0_10,
© Springer Science+Business Media, LLC 2012

archaeological careers. Therein lies the crux of a field school director, can teaching be aligned with research, can both be obtained without the sacrifice of the other? These are problems that are increased exponentially in an underwater environment.

Numerous field schools have shown that this is possible, but only with careful planning. Here a freshwater underwater archaeological field school approach is presented that includes formal and informal assessments which track both process and methodological project outcomes. The nature of the process enables this specific type of underwater archaeological field school to run in any type of environment, fresh or salt water and on any period or genre of submerged site. The eight steps outlined here include the process of designing the field school within a research project as well as the educational outcomes that are strategically developed to include students while teaching good practice and good science. These eight steps are similar to those used by businesses in project design, but at each step of the project students are involved in learning about the entire process and the specific techniques that will help them achieve success (Smith and Corbin 2010). In setting up the challenges for the students, the process opens the door to innovation and more importantly to engagement. Engendering active ownership from the project stakeholders and students alike is crucial for project and educational success.

10.2   Step 1: Establishing the Partnerships

Successful projects, whether academic or commercial, need strong, trusting partnerships. Each affiliation brings a skill that makes the project not only more viable, but increases both the chances of success and the caliber of product. The more complex the project is, the greater the need for specialists outside of archaeology. In many cases, an archaeological field school should include:

- The governing agency mandated with the underwater site's protection
- The organization leading the field school and holding the permits
- Educational institutions sending students to the field school
- Philanthropic supporters
- Local businesses
- Publishing group
- Analysis specialists

Although many of the partnerships are launched well before students arrive, it is vital that students understand the partnerships and thus the "deliverables" that each partner expects from the field school.

The wonderful advantage of partnering with governing agencies is that they have an abundance of understudied sites or areas that require exploration to determine the magnitude of submerged sites. Rivers and shorelines in lakes are excellent prospects for submerged sites. Moreover, these agencies, which vary in name from place to place, usually have a list of sites they would like to see studied – historic river landings (Sacramento River, California), abandoned ships (Tar River, North Carolina and The Murray River, Australia), aboriginal fishing (Ahjumawi Lava Springs State Park,

California), and naval encounters (Penobscot River, Maine, Lake Erie and Lake Champlain) (California State Lands Commission 1988, 2010; Babits et al. 1995; Kenderdine 1993; Ford and Switzer 1982; Foster 1999). Most importantly, partnering with an agency promotes stewardship, which is fundamental to good practice in archaeology. All of the project sites listed above fall under a governing agency's mandate and add the public as a partner to the field school experience creating a "real world" management issue as one of the field school's learning strands. Working with an agency, domestic or foreign, forces the issue of permits and defined outcomes. Applying for permits requires detailed explanations of goals, methodologies, timeframe, budgets and project outcomes, and deliverables. These real management issues impress upon students the achievement of goals in a timely manner and provide easy assessment points throughout the field school experience.

The following examples focus on the investigated shipwrecks of the Sacramento River at the point where the 1849 Gold Rush began. The 20-mile stretch from the mouth of the American River where it joins the Sacramento to just below the agricultural community of Clarksburg is the final resting place of numerous midnineteenth century ocean-going vessels, barges, riverboats, and historic landings. The Sacramento River possesses similarities to rivers around the world, varying in depth and current at different times of the year. The sediment load in the river reduces visibility to a foot at best, and zero some of the time. Dredging, fishing, and recreational traffic are a real threat to submerged cultural resources. Yet the discovery and study of shipwrecks and landings of Gold Rush California continues to illuminate a dynamic period in American history. These freshwater sites illustrate technology, trade, and the changing maritime landscape of early California. The freshwater nature of the sites provides information not often seen in saltwater environments, and the jurisdictional nature of riverine environments provides a unique arena for projects.

Possibly the most distinctive attribute of freshwater projects compared to their saltwater compatriots is the sheer number of governing agencies that require input into a project. One of the most challenging examples is the permitting structure for the Sacramento River in California. The Clarksburg shipwreck falls under the jurisdictions of the Office of Historic Preservation, State Parks, the State Lands Commission, CalTrans, the Sacramento Water Board, the Sacramento Dredging Commission, the Army Corps of Engineers, the US Coast Guard, and the National Oceanographic and Atmospheric Administration. Unfortunately, there is no overarching understanding between agencies and thus advice given by one often countermands the advice or requirements of another (Smith and Corscadden 2009). While partnerships with all of these agencies were not required, all were notified and kept informed.

In situations where the jurisdictions are complex, as at Clarksburg, it is best to bring all agencies together and let them decide who will take the lead. This process helps streamline the information shared, removes potential misunderstandings, and allows a clear timeline and deliverables. While pursuing multiple permits and keeping multiple agencies simultaneously informed requires time and strategy, their assistance provides many important resources.

Partnerships bring valuable resources to the table in expertise, personnel, equipment, public relations, and contacts. By partnering with the California State Lands

Commission and California State Parks for the Clarksburg shipwreck investigation in the Sacramento River, the field school was able to access valuable archival information, use equipment, and draw from the agencies' expertise and established networks.

The State Parks' Archives were helpful in identifying artifacts recovered from the shipwreck *LaGrange* that served as Gold Rush Sacramento's earliest jail (James and Smith 1988). The state's collections exposed students to artifacts associated with the time period while posing relevant questions regarding in-depth material culture study and collections management. As students garnered valuable experience with the artifact collections, the state gained a better understanding and more in-depth information about the collections. The longer the students worked alongside the archivists and historians, the more questions they had regarding Gold Rush California.

The institutional networks also enabled the field school to approach the river communities near the Clarksburg site for logistical assistance. The partnership with State Parks signaled to community members that the field school was important and thus deserving of their participation. The community partnerships created goodwill, supported local business, and got the students involved in stewardship promotion at a grassroots level. The skills modeled for students by strong partnerships are invaluable not only for a career in archaeology, but careers in any profession.

The partnerships between field schools and Higher Educational institutions are varied, as are the relationships between field schools and philanthropy. A sponsoring university, college, or foundation solely supports some field schools, while combinations of educational institutions and philanthropic organizations support others. A drawback of running a field school through Higher Education rather than a foundation or institute is the tuition cost for students not associated with the institution. Generally out-of-state tuition is very high and often limits the students who can attend, while foundations and institutes have a set tuition for everyone. The main drawback for institute and foundation-run field schools is the cost of the faculty. In either scenario, it is important to establish relationships with Higher Education either through direct partnership or individually through the attending students.

10.3   Step 2: Choosing Survey vs. Site Investigation

There are two main types of underwater field school whether they take place in an ocean or freshwater environment – survey or site investigation. Both types impart valuable applied learning to students. Freshwater surveys and sites have unique attributes that help students gain a wider toolkit of skills and a more varied understanding of ship construction and vessel types.

Surveys in freshwater are as varied as surveys in saltwater. River surveys more often than not deal with black water, current, agricultural pollution, and good preservation. Lake surveys often add depth and water clarity to survey variables. In most situations, riverine and lake surveys begin with sophisticated remote sensing equipment that leads to site verification. In low-visibility and black water situations, site verification requires precise search techniques in order to relocate the site. Both remote sensing and structured search patterns associated with low-visibility freshwater sites are marvelous teaching opportunities that provide applied learning for both theory and procedure. When used together, they greatly enhance archaeological outcomes. Survey techniques of all kinds engage students in conceptualization and adaptation to real environmental issues. Understanding what a site looks like remotely aids in physical identification underwater and prepares students to utilize visualization and feel versus sight alone. Plus, in black water, remote sensing often captures disarticulated fragments that, although only a matter of feet or meters away, are completely invisible to the naked eye underwater. Remote sensing images of both the *La Grange* and the Clarksburg wrecks in the Sacramento River exposed multiple hull sections at both sites. Without this information, it is likely that the smaller sections may never have been found (Hunter 1983; James 1986, 2008).

Combining remote sensing and site verification in a field school environment enables the faculty to discuss and model good practice in a low-risk situation that does not threaten the archaeological resource. The remote sensing equipment in these situations needs not be state-of-the-art to imbue students with theory and practicum. However, students need to understand the whole process from beginning to end in terms of mobilization, collecting and interpreting data, associating the data with real time and space, and reporting. In terms of mobilization, rivers add the extra excitement of snags and underwater hazards not always present in deep ocean tows. Another advantage is that snagged instruments can be recovered more easily in shallow water.

Combined surveys can be divided between remote sensing and verification, allowing lectures and training to naturally flow from locating to identifying and from technological to basic diver observation. Encouraging students to experiment with various forms of survey techniques used in locating sites and features can introduce both concepts and practices, challenging them to try multiple solutions. For example, Jackstay searches that utilize a line tied between two weighted ends of a survey may suffice on paper, but in riverine environments where trees and other debris litter the submerged banks, Jackstay searches are often difficult to implement. Spring floods in the Sacramento River carry whole trees down from the foothills – they often become wedged against anything protruding from the riverbed or out of the banks. Once a tree becomes lodged, it then becomes a dangerous attraction for others. Add entangled monofilament, fishing line, and black water and Jackstays survey methods are not practical. The same is true for circle searches in rivers where strong currents force lopsided results. However, without the experience of implementing these search techniques in varying conditions, students do not gain practical knowledge for future decision-making. Learning the realities of searching in low visibility and visualizing capabilities needed to identify sites and features are powerful tools for a student's toolkit that will only accentuate good practice and good science in more hospitable environments where visibility is high, but the shipwrecks may be less intact (Fig. 10.1).

Site investigations in the last 20 years have been primarily nonintrusive. This is true for freshwater as well as saltwater environments as the ability to conserve and then preserve collections proves complex. Waterlogged artifacts from freshwater sites demand conservation if they are to survive. As conservation is expensive, excavation is often limited in field schools. Generally, only field schools attached

Fig. 10.1 A student of the 2009 PAST Foundation field school completing a site survey dive (copyright Past Foundation)

to universities with conservation facilities excavate submerged sites. That said, diagnostic artifacts were recovered from the *LaGrange* during an investigation in 1988 and conserved, but funding for conservation was considered prior to the field study (James and Smith 1988). Without funding for conservation, it is important to give considerable thought to the site selection both for archaeological significance and the safety of the students. Thus, it is extremely important that the partnership carefully choose a site for study, prior to the field school's arrival, with consideration for numerous factors including significance, accessibility, and safety.

Considerable time is needed to plan a well-structured field school. Usually, when partnering with a governing entity, the agency has a list of known sites that are understudied. The *La Grange*, *Stirling*, *Ninus*, and *Dimond* shipwrecks in the Sacramento River were all discovered through remote sensing surveys. A sheriff found the Clarksburg wreck, which has yet to be identified by name, while searching for a car that veered off the levee road and plunged into the river (Hunter 1983; James 1986; Foster 2005). Like any other type of archaeological site, there are numerous reasons to study these known but understudied sites. Some are under threat from natural and/or human processes, though study of on-going site formation process can provide valuable information. Many sites possess potential for graduate research, and some have National Register nomination potential.

By clearly addressing the rationale for the questions asked of the site, field schools are able to educate, elucidate, and enhance valuable resources that might otherwise go unstudied. The Clarksburg site and the data recovered by the field school continues to provide important information about the threat of dredging, the ongoing site formation as the shipwreck reburies, the rich historic nature of

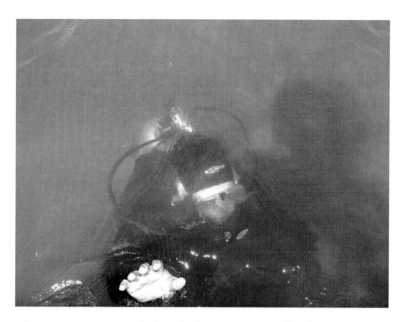

Fig. 10.2 Diver returning from a site investigation (copyright Past Foundation)

California's premier maritime highway, as well as intriguing researchers with new questions. The significance of the site will reverberate for years to come.

While partnering organizations have numerous projects that need attention, it is up to the archaeological faculty of the field school to determine suitability for access and students capabilities. For example, the Sacramento River runs faster in the summer than in the fall and has less visibility at that time, but field schools are normally held in the summer so the level of student diver certification had to be advanced while the window of diving each day was carefully considered to take full advantage of slack tide. Fundamental to a successful underwater field school is the match in capabilities of the student as a diver and the conditions in which he or she is asked to work. Archaeological field schools in black water riverine environments generally require as much if not more attention to safety as those field schools executed in dynamic surf environments. The accessibility and safety of the shipwreck sites in the Sacramento River were therefore carefully assessed before choosing the appropriate location for the field school.

In the best of situations, field schools can combine attributes of one site with those of another to create a fully robust field school experience that covers material culture, underwater techniques, theory, conservation, site management, and stewardship. In the case of the Clarksburg shipwreck field school, the students studied the artifact collection of *La Grange* and the collateral artifacts linked to persons associated with the ship or jail then documented the Clarksburg shipwreck underwater. The ability to combine studies reflects both the open communication between partners and the clear goals of the field school (Fig. 10.2).

10.4   Step 3: Determining the Research Questions

Once the significance of the site or sites that constitute the focus of the field school is identified, a brainstorm of relevant questions begins. Each partner will have questions. In the case of the Clarksburg shipwreck, both State Parks and the Army Corps of Engineers wanted to document and date the ship construction details in an effort to help future archaeologists better ascertain the historic value of emerging sites along the river. Historians wanted to know the ship's identity. The State Lands Commission wanted to know if the condition of the splayed hull was recent or historic in order to better understand the impact of dredging on the river. The State Archivists wanted to better understand the archaeological collections from the Sacramento area that are alleged to have associations with *La Grange* (Foster et al. 2007).

Each question was valid, but in the context of a field school had to be carefully orchestrated to acquire answers. As with all archaeological projects, permitting and funding constraints shape how fully the queries can be answered. The questions were ultimately organized to correspond with the structured flow of the field school. The artifacts were studied first, supplying the students with a sound foundation in the material culture of the Gold Rush and providing the State Archives with a more complete understanding of the associated materials. Documentation of the shipwreck took into account the hull construction, the site formation changes, and the site environs, instructing the students in techniques and strategies. The combined information supplied the diagnostic information that ocean-going hulls journeyed up the Sacramento River prior to the advent of hydraulic mining in 1852 (James and Smith 1988). The documentation also shed light on the dynamic nature of the reburial of the shipwreck. Although indicators suggest the ship is reburying itself, the information is too general to predict time spans with any accuracy. The hull construction documentation, wood samples, and copper samples are important clues that may help historians identify the ship (Smith et al. 2008; Smith and Corscadden 2009). By the end of the Clarksburg field school, students collected information for three of the four research questions, with only the limitations of time preventing study of the dredging impact on the site.

At each step in the project, the needs of educating the students, keeping everyone safe, and answering the research questions must be regularly monitored. Transparency of the process helped the students better understand the intricacies of research design, practical application, safety, permit restrictions, and partnership responsibilities.

10.5   Step 4: Aligning Field Schools Schedules to Learning

Once the research questions are defined, the teaching topics, methodologies, and schedules are timetabled. Alignment of research and learning helps develop efficient schedules as well as define clear-cut budgets.

Fig. 10.3 Evidence of the maritime cultural landscape of the Sacramento River documented by the 2009 PAST Foundation field school (copyright Past Foundation)

Time, function, and location of the site or survey clearly define many of the learning topics within a field school. In short, the material culture, the ship construction, underwater techniques, the conservation, and site management covered within the course topics must fit the site or survey area. In the example of the Clarksburg shipwreck, understanding the time period and location of the wreck was helpful in combining the material culture of the *La Grange* and the surrounding cultural landscape of the Sacramento River as a maritime highway (Smith and Corscadden 2009). The two sites fall within a 5-year time span and are located within twelve miles of one another. As part of the daily ride back to the marina, students took GPS locations of old pilings and remnant brush landings along the embankments. These locations were mapped in an effort to better understand the context of the Clarksburg wreck and place it within the historic landings system (Fig. 10.3).

An archaeological project run with a professional crew tends to have well-defined methods, procedures, and protocols, whereas field school methodologies are more variable in order to expose students to differing techniques so that they can assess and choose appropriate methods. The level of field school, whether introductory or advanced, defines the depth to which the research questions are likely to be answered and the variety of methods introduced or presented to students. The Sacramento River is relatively shallow today due to silt run-off from agriculture and the sediment where snags are not present is either sandy or silty. The employment of several survey and measuring techniques during the underwater work exposes the limitations and advantages of differing techniques. The ability and willingness to discard one methodology for another helps students stay open to possibilities and be better prepared for unseen events. Opening floodgates upriver can raise the water level and dramatically increase current. Being able to switch from one methodology to another helps students to continue working with minimal disruption.

Aligning the tasks and learning in ways that provide strong foundations for critical thinking helps expedite students' learning and enables them to readily see their own growth and success (PAST 2009). By establishing a learning path or schedule beginning with material culture, before the students ever set foot in the water, allows the faculty to clearly differentiate the archaeological goals from the tools we use to achieve them. Moreover, understanding material culture helps with rapid diagnostic identification underwater, saving time at critical points in documenting a site.

10.6   Step 5: Linking Learning to Lectures

Good interpretation, like good practice, needs to be modeled, but without foundational knowledge it is difficult to do either. Short morning lectures that impart information are one key method for instilling information and methodological theory, providing students with an expanding toolkit of knowledge. The process proven quite successful for the PAST Foundation is a short morning lecture or discussion followed by hands-on experience which is consolidated through a debriefing discussion. This can then lead on to more hands-on experience, culminating in an evening lecture.

This is a common project-based learning structure that is tied to the goals of the project and learning needs of the students (Benett 2007). At the Clarksburg project, students enjoyed guest lectures on site management, California history, site formation in California Rivers, and conservation along with lectures on material culture, ship construction, cultural landscape, and dive safety. Each of the lectures was paired with practical experience. The lectures were scheduled to help build on existing knowledge that culminated in an understanding of the project's scope from the creation of partnerships and research design through the delivery of a completed report and presentation of learning.

10.7   Step 6: Linking Learning to Real Outcomes

Authentic learning, solving real issues through practical applications, is an essential component of a successful field school. Besides being the most effective pedagogically, the students walk away with a fulfilling educational experience that inspires them to become good stewards of cultural heritage. A successful field school has the flexibility to allow the students create their own learning paths within a given structure. This approach requires that the faculty become learning guides rather than revered sages. Individualized paths involve success and failure but ultimately triumph. Learning requires brainstorming, solution building, testing, and evaluation – simply being told is rarely enough. Using these design principles or scientific methods, students can attack the challenges of the research questions and through systematic processes find answers that have an authentic grounding. The research questions posed by the partners were intended to enhance understanding and enable better management of the cultural resources of California. The outcomes of the questions have the real potential to affect local, regional, and statewide decision-making. The added weight of authenticity helps ground the students and impress upon them the seriousness of the undertaking. There was no time-filling assignments at the Clarksburg field school, although many tasks were tedious.

The final step is to associate the student learning to substantive outcomes. Involving the students in the final leg of a project instills responsibility and assists in synthesizing collected data. How do you know that the posed research questions have been answered if the data are not synthesized into a presentable form? How do you know where the gaps are in your data collection? What better assessment of recommendations for future research than the presentation of collected data flaws and all? In the synthesis of data for the Clarksburg field school, students realized that no conclusive data were gathered to address the question of dredging damage to the site, even though discussion and numerous informal observations regularly tackled the issue.

10.8   Step 7: Presentation of Learning

Presentation of learning comes in many forms and is more than simply the regurgitation of knowledge, and it allows the students to demonstrate that they have collected, interpreted, and synthesized data to be presented through a variety of formats. This is an important stage in the education process, where the field school faculty helps students find their strongest means of expression so that everyone contributes to the final project. A presentation of learning is a means for every student voice to be heard.

The learning structure for team building in the project must begin at the start of the field school with a clear understanding and willingness to explore team member strengths. Not every student will be a superior diver or an artist, or top-flight researcher, but together the overall team may possess all these qualities. Learning to depend on one another builds outcomes that far exceed the mere sum of the parts. Exploring the characteristics of what varied talents make up a strong team helps students better understand the distinct roles within the group. Understanding diverse team member strengths also helps groups strategize efficient ways to tackle problems and multitask a team. These are skills that might not be considered strictly archaeological in nature, but directly affect the outcome of a successful project, and are important transferable skills that the students acquire.

Presentations of learning may be a site map, historical research, clear photography, good writing, clean database, documentary film, or report layout. The entire team will have contributed to the completed product. Making the final presentation to the project partners reinforces the culminating task of a project and brings the project full circle. In the case of Clarksburg, the first two field school reports drove the succeeding field school goals and generated graduate studies as well as changes in cultural resource management. The documentary podcast and digital presentations have been viewed by a wide audience and also presented at numerous conferences.

10.9   Step 8: Postproject Brainstorm

No matter how successful a field school may appear upon completion, it is essential to evaluate how the program could be improved; everything has the potential to grow and evolve. Methodologies and lecture material should be kept current, while the success of the partnerships and the ability of the field school to achieve the goals set need to be evaluated annually by all stakeholders. It is also important to recognize unexpected successes that might be incorporated into future field schools. For example, the addition of looking at the historic river landings to and from the site each day added a contextual understanding that the students thoroughly enjoyed. The postfield school evaluation is a time to critically assess the success of the project budgeting and expense management; without good fiscal oversight, even the best field school will flounder. Such reflections also enable all the partners to revisit the research questions and decide well in advance if further work is needed, which can potentially define future field schools or graduate research programs.

10.10   Conclusion

By making sure that the eight steps listed above are incorporated into the planning and execution of field schools, whether they are in rivers, lakes, oceans, or on land, helps ensure that students receive grounding of knowledge, theory, and good field practice. Helping to structure knowledge for the students in a logical way, while encouraging them to experiment and question, ensures that they develop vital critical thinking necessary for their personal development. Field schools play an integral role in the holistic education of archaeologists, and exposing students to an array of underwater site types, including a variety of site environments such as those in freshwater, broadens student knowledge bases and bolsters their technical tool kits. In this way, the next generation of reflexive and adaptable underwater archaeologists will be nurtured.

References

Babits, L. E., Morris, J., & Corbin, A. (1995). *A survey of the North Shore Pamlico River: Bath creek to Wade's point*. Greenville: East Carolina University [manuscript].

Benett, S. (2007). *That workshop book: New systems and structures for classrooms that read, write and think*. Portsmouth: Heineman Press.

California State Lands Commission. (1988). *A map and record investigation of historical sites and shipwrecks along the Sacramento River*. Sacramento: State Lands Commission.

California State Lands Commission. (2010). *Shipwreck database*. Retrieved April 25, 2011, from http://shipwrecks.slc.ca.gov/ShipwrecksDatabase/Shipwrecks_Database

Ford, B., & Switzer, D. (1982). *Underwater dig*. New York: William Morrow.

Foster, J. (1999). *Archaeological investigations of the Ahjumawi Lava springs fish traps*. California State Parks.

Foster, J. (2005). *Clarksburg shipwreck reconnaissance*. Sacramento: California State Parks.

Foster, J. et al (2007). *Archaeological survey of cultural resources: Sacramento River*. California State Parks.

Foundation, P. A. S. T. (2009). *Summer bridge programs, 2009*. Columbus: PAST Foundation.

Hunter, J. (1983). *Investigation of submerged cultural resources survey, Sacramento River, California*. Sacramento: Sacramento Housing and Redevelopment Agency [manuscript].

James, S. R. (1986). *Submerged cultural resources survey, Sacramento Embarcadero, Sacramento, California*. Sacramento: Sacramento Housing and Redevelopment Agency [manuscript].

James, S., & Smith, S. (1988). *La Grange, A California Gold Rush Legacy*. California State Parks.

James, S. et al (2008). *Investigation of Submerged Cultural Resources: Sacramento River Clarksburg*. United States Army Corp of Engineers, Sacramento Region.

Kenderdine, S. (1993). *Historic shipping on the River Murray NEW/Vic*. Paramatta: Department of Planning [manuscript].

Pyburn, K. A., & Wilkes, R. R. (1995). Responsible archaeology is applied anthropology. In M. J. Lynott & A. Wylie (Eds.), *Ethics in American archaeology: Challenges for the 1990s* (pp. 71–76). Washington: Society for American Archaeology Special Report.

Smith, S., & Corbin, A. (2010). *Problems, programs, projects: Designing transdisciplinary problem/project-based learning*. Columbus: PAST Foundation.

Smith, S., & Corscadden, A. (2009). *The Clarksburg Shipwreck archaeological field report 2009*. Sacramento: California State Parks [manuscript].

Smith, S., Corbin, A., & Corscadden, A. (2008). *The Clarksburg Shipwreck archaeological field report; 2007–2008*. Sacramento: California State Parks [manuscript].

Part IV
Non-Excavation

Chapter 11
Pompeii Food and Drink Project

Betty Jo Mayeske, Robert I. Curtis, and Benedict Lowe

11.1 Introduction

The goal of the Pompeii Food and Drink Project is to conduct a noninvasive comprehensive surface investigation of all structures in the Roman city of Pompeii and select neighboring structures to discover any overall patterns of daily life associated with food and drink. To accomplish this goal, we established four research objectives: 1. To identify rooms, spaces, and features associated with all aspects of food and drink, including production, processing, storage, preparation, selling, and consumption. 2. To document the rooms, spaces, and features of structures 3. To organize and preserve the above information in a database and in print forms, and 4. To employ a geographic information system (GIS) to disclose patterns of living associated with food and drink in Pompeii and outside the city walls. The project is not associated with a university. Rather the Project Researchers, staff members, and team member/volunteers plan, conduct, and pay for the project. This chapter details: planning and preparation in the USA; acquiring Italian authorizations; recruitment of team members; finances and budgeting; accommodations in modern Pompei; the on-site research and documentation experience; post field work; and a summary of what we have learned after ten years of research in ancient Pompeii.

B.J. Mayeske (✉)
History and Humanities, University of Maryland University College,
Adelphi, MD 20783-8075, USA
e-mail: BMayeske@faculty.umuc.edu

R.I. Curtis
Department of Classics, University of Georgia, University of Georgia Park Hall,
Athens, GA 30602-6203, USA

B. Lowe
Department of History and Area Studies, University of Aarhus, 16 Bartholins Alle,
Aarhus C, DK 8000, Denmark

H. Mytum (ed.), *Global Perspectives on Archaeological Field Schools:*
Constructions of Knowledge and Experience, DOI 10.1007/978-1-4614-0433-0_11,
© Springer Science+Business Media, LLC 2012

Table 11.1 Pompeii food and drink project assumptions

The project will limit analysis of food and drink to all structures, including private houses, shops, public buildings, and temples, or rooms within these structures inside the city of Pompeii and selected villas and tombs outside the city, as they functioned in AD 79
Analysis of the physical remains available for study today will require not only grounding in previous excavation reports and subsequent publications but also on-site close examination by PRs
The vast amount of information collected and needing to by analyzed will require a large and sophisticated database
A project of this complexity will require many individuals, some already possessing desirable specific skills, while others will need training
While the project will focus on research, educating team members on Roman culture and preservation of an endangered site will receive a high priority
Publication of data and its analysis and providing access to the data by future researchers will be priorities of the PRs
The ancient city covers more than 150 acres and contains over 1,200 structures, so a project of this scope requires many years to complete
A long-term project of this size and complexity requires an outside source of funding

The goal of the Pompeii Food and Drink Project is to conduct a noninvasive comprehensive surface investigation of all structures in the Roman city of Pompeii and select neighboring structures to discover any overall patterns of daily life associated with food and drink. To accomplish this goal, we established four research objectives:

1. To identify rooms (part of a structure enclosed by walls), spaces (an open public or private area), and features (distinctive or characteristic objects within a room) associated with all aspects of food and drink, including production, processing, storage, preparation, selling and consumption.
2. To document the rooms, spaces, and features of structures by written descriptions, measurements, sketches, formal drawings, photographs, and video records.
3. To organize and to preserve the above information in a database and in print form.
4. To employ a geographic information system (GIS) to disclose patterns of living associated with food and drink in Pompeii and outside the city walls.

Nine dominant assumptions ground the concept of the project, and these are listed in Table 11.1. Although the primary focus of Pompeii Food and Drink is research, the PRs recognize the obligation to provide team members with a valuable learning experience in exchange for their financial contribution and labor. Our educational goal is to contextualize the focus of our research project, to introduce team members to other scholars working in Pompeii and their projects, and to increase the knowledge and appreciation of Roman culture generally and of life in ancient Pompeii specifically. In pursuit of these objectives, we have instituted four educational programs.

Before the season begins, each team member receives a Project Briefing Book that includes not only advice about what to bring with them and what to expect from living and working in Italy, but also introductory information about the Project

Fig. 11.1 Evening lecture by the staff member at the motel

and its goals and about ancient Roman culture, including a vocabulary list of archaeological terms and translations of relevant passages from select ancient authors. In addition, on the first full day of each summer season, a PR conducts a guided, limited on-site tour of Pompeii. The purpose is to familiarize team members with the layout of the ancient city, to acquaint them with the types of structures in which they will be working, and to introduce them to the vocabulary particular to our research project. This latter goal is especially important since many team members possess neither prior knowledge of Pompeii, Roman history, Latin language, or ancient architecture nor archaeological experience. For those joining the project in the second or third week, a Staff member conducts a tour in the early morning of the first day. In addition, PRs throughout the time at Pompeii encourage individuals to pursue discussions on any topic as occasions arise.

On three or four evenings each week just prior to dinner, PRs and Staff present poolside lectures (Fig. 11.1). Subjects vary but are chosen either to augment the research focus or to introduce new, but related, topics. Themes include Roman architecture, Roman history, dining in the Roman world, archaeological drawing, religion, and the Pompeii Food and Drink database. In addition, as opportunities arise, we invite other scholars already in the area to talk to the group about their work.

In 2008, we instituted a series of on-site morning lectures. This innovation arose from the need to occupy team members in the early morning, while PRs made initial investigation of the structures identified for study on that day. Staff personnel present talks that focus narrowly on particular structures or limited topics. Where possible, these presentations take place on the site of the subject matter. Themes have included the Temples of Isis and of Apollo, the forum, baths, theaters, commercial shops, vineyards, and wall painting, among others.

PRs encourage team members to visit other sites around Pompeii during their free time, particularly on weekends. Indeed, where to go and how to get there form the topic of a specific evening lecture. In addition, during the second week, the Project sponsors a group visit to the Villa Regina at Boscoreale located about a mile distant from Pompeii. This site directly relates to Project research in that it includes a farm house destroyed by the eruption of Mt. Vesuvius and a small museum that emphasizes artifacts relating to food and drink. In 2008, on a day that a custodian strike in the ancient city of Pompeii prevented the group from entering the *scavi*, the project sponsored a site visit to Villa A (also called the Villa of Poppaea) at nearby Oplontis, a site that otherwise the Project highly encourages team members to visit on their own. And finally, the Project tries to arrange with the Italian authorities for special visits to sites otherwise inaccessible to the general public.

In 2000, Mayeske approached Earthwatch, a nonprofit, tax-exempt institute that provides project funding. Upon their approval of our application, beginning in 2001 and for the following 2 years, the PRs assembled staff, made arrangements for accommodations, obtained permission to work in Pompeii from the archaeological Superintendency, and acquired equipment, while Earthwatch recruited volunteers and set the price of their participation. Under Earthwatch, participants were denoted as volunteers; we now call our participants, more accurately we think, team members, and they are referred to in this way in this chapter.

For their work, Earthwatch retained a percentage of volunteer fees; the project used the remainder to fund accommodations, to purchase needed equipment, and to support PR travel. The Project worked in Pompeii for 4 weeks each season, during which time Earthwatch provided volunteers for two 2-week periods. This arrangement worked well until 2003 when Earthwatch's priority shifted to projects more specifically environmental in focus. Following a year of reorganization and planning, Pompeii Food and Drink in 2005 began its existence as an independent project in which it recruits its own team members and funds its work entirely from participant fees. Finding the 4-week season too exhausting and too expensive, the Project reduced the length of each season to 3 weeks, usually late June to mid-July.

11.2   Planning and Preparation

Planning and preparation preceding actual work in the field begins soon after returning from the previous summer's work. PRs must create a season schedule based upon two factors, where work terminated at the end of the previous season and an estimation, grounded on past experience, of how many structures can be analyzed and documented in fifteen workdays. Backup plans for unexpected problems, such as labor strikes, are particularly important. In early August, the Director sends this schedule to Staff, while the responsible Researcher readies materials, such as house plans and pertinent bibliography, for the rest to use to analyze structures in the field. The schedule also includes preliminary arrangement for the lecture series, both those given by PRs and Staff and those to be given by special guests.

For planning purposes, PRs and Staff meet twice each year, usually in late fall and late spring, at the home of the Director. In addition, we also try to gather at the annual meeting of the Archaeological Institute of America that usually convenes in January of each year. PR maintains frequent contact by phone and email during the year. Four major requirements dominate the planning phase: obtaining permission to work in Pompeii, confirming hotel accommodations, recruiting team members, and forming the Staff.

11.2.1 Authorization

Authorization to work in Pompeii must be requested and approved each year. In early November, the Director sends a letter to the Soprintendenza Archaeologica di Pompei requesting permission to work in the ancient city. This letter specifies the particular regions and *insulae* in the city where we wish to work. Upon approval, in late spring, this request is followed by a list of all individuals participating in the Project. The Soprintendenza uses this list to issue entrance passes.

11.2.2 Recruitment

Predicting the appropriate number of participants remains an area of difficulty. On the one hand, the project needs a minimum number to remain financially viable, while, on the other hand, available hotel space and the amount of on-site work limit the number of team members that we can accommodate. Nothing is more conducive to a fractious, disorganized, and inefficient season than too many underemployed team members.

Team members need to have no prior experience to be eligible to participate, although we have set a minimum age of 18, and have accepted applicants up to 82 years old (Fig. 11.2). Our major criterion for acceptance is the desire to work and to learn. In 2005, when we assumed full responsibility for recruitment, we targeted appropriate professional and general publications-specific groups, such as high school Latin teachers, college Classics majors, local AIA members, and food and drink aficionados. In addition, through public lectures and classroom presentations, we encouraged all interested students to participate. These efforts have not proven particularly successful. While we continue to attract individuals from these groups, many of our recruits are veteran team members or those who learn of us by word of mouth from previous participants or, especially, through our website. The website is

Fig. 11.2 Group photograph showing age range of a typical team

updated following each season in Pompeii and throughout the year, as required, being careful to keep it accurate, appealing, and helpful. Where appropriate and when requested, we cooperate with various universities in providing academic credit for participation.

Since about one-third of our team members return in subsequent summers, during the intervening year we try to maintain close contact with them to encourage their return. Especially helpful has been sending to them a copy of the field report for their particular season. We also offer to returnees a reduction in the weekly rate and, in recognition of their experience, we allow them to sign up for 1 week, if they so desire. New members must register for a minimum of 2 weeks.

In addition to team members, the project needs experienced and skilled Staff members. We have been fortunate to enjoy the benefits of a trained and committed Staff throughout the Project's duration. Recent seasons have seen a significant retention of Staff members from one season to the next in contrast to a rapid turnover experience during the early years of the project. Several of our Staff, for instance, have worked with the project for 7 or more years, a few even for the entirety of the time. This has provided important advantages: continuity, expertise, and the ability and confidence of PRs to delegate increasing responsibility to Staff members. That being said, communication among PRs, Staff, and team members needs to be close and consistent. Discordant opinions expressed by PRs and Staff can injure morale of all members of the Project, initiate friction and confusion, and create an impression on team members of a lack of professionalism.

The Director, in consultation with the PRs, identifies talented, energetic, and specially trained individuals for leadership roles during the planning phase in the

United States, fieldwork in the ancient city, and database entry in the motel. Specific Staff duties include website administrator, field manager, artists, photographers, and database managers and developers. Staff members are frequently Food and Drink Project-experienced and many have worked as archaeologists, anthropologists, architects, photographers, editors, Information System developers, and field archaeology artists. In return for their contributions, they receive free room and board. Based upon team member enrollments, no later than January the Director sends selected Staff a formal letter requesting their services for the following summer. Staff are expected to cooperate with PRs and fellow Staff members under all circumstances, to train and oversee those assigned to their particular team in a courteous and productive manner and to monitor the needs and requirements of all team members. In regard to the latter expectation, Staff personnel can effectively function as a conduit of communication between PRs and team members. They are especially useful in quickly identifying problems and relaying the concerns to PRs for rapid resolution. This function is crucial in maintaining an efficient and pleasant work environment. Besides team responsibilities related to Project work, Staff members prepare and present short lectures in the field and in the evenings at the motel. And, finally, Staff have the responsibility to prepare an end-of-the-season field report for their team.

11.2.3 Finance and Budget

The Director and one Staff Member manage the financial status of the Project. Upon separation from Earthwatch, for tax purposes the Director established a Limited Liability Corporation (LLC) with the state of Maryland and set up a business bank account to obtain project checks, to wire funds and to hold deposits. The responsible Researcher prepares a budget based upon the anticipated Euro-dollar exchange rate and the financial requirements for an estimated ten PR and Staff personnel and from 16 to 18 team members per week.

The development and ongoing revision of the budget is one of the most important responsibilities. PRs must generate sufficient funds to meet anticipated expenses and to allow for unexpected costs. Throughout the year as recruitment and confirmation of participation develop, they also closely monitor the relationship between anticipated income, based on the number of team members and projected costs. This is particularly crucial in periods where exchange rates can vary widely. A preliminary budget created in the early fall undergoes periodic revision as exchange rates fluctuate over the months. It receives its final form in the month preceding the start of a new season.

Between 2001 and 2003, Earthwatch determined volunteer numbers, set charges, and decided what percentage of that income the project would receive. Once we left the umbrella of Earthwatch after the 2003 season, we determined our own costs and developed the necessary income. Each year, the Project Director sets a weekly price for team members based on several factors. The Project Director, in coordination with other PRs, determines the minimum and maximum number of volunteers needed to perform Project tasks. Since we need to create a budget based on projected levels of participation, we require of each potential participant a nonrefundable deposit. The Staff member who oversees recruitment reminds all enrolled team members when final payment is due. The PR responsible for developing the budget, working closely with the Project Director, then calculates the cost of accommodations (rooms with air conditioning and one-half board) and anticipates other project expenses, such as new and replacement equipment, rental car, and staff expenditures arising from room cost reductions. These factors provide a general idea of how many team members we need to recruit to cover all costs. If projected costs exceed estimated income, as initially appeared likely for our 2008 season, some elements usually included, such as lunch supplement or special site trips, may have to be curtailed or canceled, or a surcharge added.

11.2.4 Accommodations

Some projects at Pompeii have chosen to live in camping sites that, while they are close to the work site and are relatively inexpensive, have limited facilities and amenities. For short periods of time this may prove sufficient, but for extended periods lack of comforts can quickly affect morale and reduce efficiency. Our ultimate choice, the Villa dei Misteri Motel, was not a difficult one. The location is about one-quarter mile from the entrance to the *scavi* (excavations) and so entails only a short walk each way. It is close enough for team members and Staff, if desired, to reenter the *scavi* in the late afternoon after work to pursue individual research or to tour parts of the ancient city not covered during the work period. Additionally, motel management provides storage facilities where we can leave equipment between summer seasons.

The motel has a large number of comfortable rooms of ample size for double or triple occupancy, most of which are air-conditioned. Amenities rarely offered by other motels in the area include a swimming pool, restaurant, bar, and large meeting room. Consequently, the motel provides a convenient and pleasant environment both to enter data into the project database in the afternoons, to hold group meetings, and to relax after work. The chef daily provides two meals distinguished for their variety and high quality. The management and staff have excellent English language skills that facilitate communication with team members most of whom are not proficient in the Italian language. The general upkeep of the motel is excellent and the management has initiated continuous improvement of facilities over the years. The only downside is cost. Room costs are probably higher than that charged by other field schools at Pompeii, but this is more than offset by the advantages noted above and by consistently high participant morale, particularly among older individuals who make up a significant portion of participants.

The responsible PR coordinates with motel management regarding room reservations for Staff and team members and acts as the contact point to respond to any problems that may arise regarding accommodations. In prior years, the PR initiated planning in early fall, but since 2006 he has concluded preliminary arrangements for the next season before departing Pompeii in July. The initial estimate of the number of rooms required derives from the estimated number of Staff plus a full complement of 16 to 18 team members per week. In addition, he reserves a few rooms for early arrival and late departure of PRs and Staff. Once the recruiting and signing up of team members have concluded, usually in early May, based on a final count of team members and Staff, the PR confirms with the motel the number of rooms required for the season. This final count should be done as early as possible in case additional rooms are needed and, should any reserved rooms become unnecessary, to allow the management to rent them.

The same PR who books the rooms also has the important task of assigning roommates for each room. Doing this properly helps to ensure a pleasant experience for everyone. Although rooms in the motel can accommodate triple occupancy, unless so requested, rooms are assigned on the basis of double occupancy. Requests by individuals for a particular roommate receive priority. In the absence of specific wishes, the PR establishes criteria for assigning roommates, keeping in mind, as far as possible, the goal of assigning compatible individuals as roommates. Success in this matter goes a long way to promote a positive experience for all participants. Among the most important criteria for assigning roommates is family relationship, such as husband and wife, parent and child, and siblings. Same gender roommates are always maintained, and, to the extent possible, assignments are age-sensitive.

On arrival in Pompeii, the Director, in her opening remarks, stresses that team members, once assigned to rooms, must be considerate of their roommates. This is particularly important for a number of practical reasons. Since in our case the motel has only one key per room, it is important when both individuals are out of the room that the key be left at the front desk. This allows a roommate who needs to return to the room without the presence of the other roommate to do so and also permits the cleaning staff to enter the room to perform their tasks when both roommates are absent.

Room assignments, however carefully crafted according to reasonable criteria, do not guarantee success in every instance. Personal characteristics and habits of roommates do not always mesh. While most team members have been compliant, tolerant, and sensitive to the needs of their roommates, instances have arisen that required reassignment of roommates. Attentiveness to potential conflicts, a quick response in attempting to resolve them, and flexibility in accommodating legitimate complaints can avert serious morale problems. PRs should move quickly to resolve conflicts as they arise and, if a mutual resolution is not found, to respond to requests for room assignment changes. This requires the utmost patience, interpersonal skills, and sometimes ingenuity.

As important as are the accommodations themselves, the development and maintenance of a good relationship between field school personnel and motel management and staff are crucial for success of any long-term project. This rapport ensures continuity of accommodations from year to year and encourages a relaxed atmosphere for all. The Villa dei Misteri Motel is a family-owned and operated establishment and many of the extended family reside on the premises and share much of the facility with motel guests. The intimacy of the relationship between Project personnel and motel management and staff has increased over the years until now a close friendship exists. This bond was not achieved overnight but developed from a concerted effort by both parties to work closely together for their mutual benefit. Confidence grew as the Project returned each summer, as personnel in both groups remained generally consistent, as the trust level increased with growing familiarity, and with the recognition on the part of management that Project members were sincerely trying to accommodate their requirements and with the Project's recognition of management's attentiveness to its needs. It is also important to consider the role motel staff play in the morale of Project participants during their time spent out of the *scavi*. We are at pains to show our appreciation not only by promoting good interpersonal relations, but also by rewarding their work for us with appropriate tips at the end of each week.

11.3   The On-Site Experience

Working in Pompeii presents several unique problems stemming from its fragile physical state and the importance of the ancient city as an active archaeological site and major tourist destination. Our work begins in the early morning with the opening of the site to the public. Although some areas of the site are readily accessible, many parts are closed. While we have permission from the Soprintendenza to work in the site, in practice access to individual houses depends upon the on-site custodians. On the first day of the season, the Project Director secures entrance passes for all Project members from the office of the custodian. On occasion this has taken a considerable amount of time during which team members were unoccupied. This problem was resolved by conducting a series of lectures at the motel introducing team members to Pompeii and culminating in a tour of the site. Indeed, filling dead time is a problem needing close attention since failure to do so can affect team member morale.

In most cases, custodians quickly honor our request to enter a particular structure, but unforeseeable problems do arise from time to time. In these cases, patience and flexibility are necessary virtues. At the beginning of each morning, two PRs go to the office of the custodians with a list of structures needing to be entered that day. Even planning ahead to identify particular structures for which access is needed, however, does not guarantee an efficient process. A considerable amount of time to gain entry sometimes becomes necessary when the structure has not been opened for many years and either the keys are missing or the lock has rusted shut and needs to be cut. At these times, having identified an alternate list of properties is important. In addition, one of our concerns during the early years of the project was to keep team members occupied during these delays. In 2008, we alleviated this problem by scheduling on-site Staff lectures on a variety of themes relating to Pompeii and food and drink. This provided time for the PRs to gain access to the structures and to carry out the initial identification of rooms, spaces, and features of interest.

Difficult access to properties can take other forms as well. Contemporaneous work by other archaeological projects in structures of interest to us, for example, can delay or even prevent our access. While we have permission to work throughout the site, coordination with other archaeological teams is essential to ensure that each can carry out research without detriment to the other. Where an accommodation has proven impossible, we have delayed our work to a later time or, on rare occasion, to the following summer. Structural instability of a property can sometimes render part or even the entire structure unsafe to enter. The earthquake of 1980, for example, caused extensive damage to the city some of which today remains unrestored. Consequently, as happened during the 2011 season, access is at times restricted to only a small group or for only a limited time period, or even denied altogether. Since it is essential that we gain access to all structures in the city, we must request from the Soprintendenza permission to enter that particular structure the following year. Happily, the Soprintendenza has been supportive of our needs and we have never ultimately been unable to enter a house.

Once we have entered a property we divide our research among four teams of Staff and team members. The PRs lead team one that initially identifies spaces and features relating to food and drink and records the information on a feature sheet (Fig. 11.3). In addition, a Staff member makes a video record of each feature. Individual Staff Members lead other teams in recording additional information for each feature. Team two photographs all the features and rooms identified by team one (Fig. 11.4), while team three draws and maps them (Fig. 11.5).

As we gained more experience in identifying features and particularly as we encountered unanticipated situations on-site, we have had continually to re-evaluate our research methods and data organization. Responses to changing technology, such as improved digital camera and software capability, have also led to adjustments in how we gather data and record them in the database, a process that is no doubt to continue. Because of the heat, our on-site work finishes in early afternoon, after which team members have a few hours free to relax and enjoy the amenities of the motel. Work resumes in midafternoon when team members and Staff enter into the database the data gathered during the morning. In addition, individual teams are responsible for cataloging and digitizing photographs and drawings.

The efficient completion of our research within the constraints imposed by the nature of the site outlined above requires considerable organization in regard both to Staff and team members. Since 2002, a Staff person has functioned as field manager to coordinate activities of all three teams to ensure that they meet the needs of our research and the requirements of both the Soprintendenza and the custodians. In 2008, we instituted a Quality Assurance Inspector, a Staff member who reviews descriptions recorded on feature sheets to ensure that they contain all necessary information before team one leaves a structure. Assignment of team members to particular teams has also been a perennial problem, as yet not adequately resolved. On the one hand, assigning team members to a particular team for the entire 3 weeks of the project enables them to gain experience and camaraderie. On the other hand, some team members have expressed the desire to try their hand at all aspects of our research.

Fig. 11.3 Team one identifying features by room at Pompeii that are associated with food and drink

So, during the 2008 season we rotated team members from team to team on a weekly basis. Our greatest concern is to balance the team members' desire for a variety of work experiences with the anticipated loss of time it will take to retrain that person in a new job each week. Experience will tell us if we can find that delicate balance.

Communication is absolutely vital to coordinate the activities of up to thirty Staff and team members. Word of mouth is not sufficient. Since 2006 the responsible Staff member has posted a daily schedule and team assignment list on a notice board in the motel lobby. The schedule includes a full daily itinerary, including the time of entry and return from the site, work assignments for the afternoon, and evening lectures or site visits. The assignment list identifies which team member has been assigned to a particular team for the coming week and gives their work assignments for the afternoon. The responsible Staff member assigns afternoon work to team members on a day on/day off basis. This ensures that team members know in advance where they must be on a particular afternoon and what they will be doing.

Fig. 11.4 Team two photographing relevant features at Pompeii that are associated with food and drink

Fig. 11.5 Team three drawing the relevant features at Pompeii that are associated with food and drink

This enables them to plan ahead for other activities. In addition to the daily schedule, postings include relevant local information such as bus and train times.

Each Sunday evening, PRs, Staff, and team members meet before dinner. At this time, the Director conducts general introductions, makes necessary announcements, and discusses pertinent issues, such as health and safety concerns. In addition, PRs and Staff meet twice weekly to discuss progress of on-site work and to identify any problems that have arisen. Addressing quickly and efficiently team member dissatisfaction is a continuing concern. Since the 2008 season, one designated Staff member monitors closely team member morale to identify and to seek to resolve any difficulties arising during the course of the season. That Staff member also functions as the liaison between team members and PRs to ensure two-way communication and quick response to concerns. We have found that a single discontented team member can disrupt the balance of the entire season and the sooner potential difficulties are identified and resolved the better.

Team members constitute the core of our Project; without them it could not function. During our years under Earthwatch, participants came from a wide variety of backgrounds; some already had an interest in the ancient world, but many joined the Project through their interest in the goals of Earthwatch. They brought varying expectations. While many were motivated by a serious interest in the Project and a desire to work, others saw the season as an opportunity to explore Pompeii and the surrounding region, developing a "tour" mentality toward their participation. We were able to offset much of this lack of seriousness by early and clearly communicating what we expected of participants and providing a daily schedule of work assignments and related activities. In essence, the more professional we appear, the more professionally team members will respond. After 2003, while the range of backgrounds remained wide, we attracted an increasing number of team members with an interest in ancient studies, including many seeking bachelor and advanced degrees in archaeology or a related field. This increasing core of expertise helps to avoid many of the pitfalls of working in Pompeii and engenders camaraderie within the project.

Safety is a major concern because of the fragile physical state of the archaeological site and the day-to-day environmental hazards of working in southern Italy in late June and July. We include health and safety procedures in the Briefing Book, covering such basics as the need for sunscreen, hydration, and the wearing of suitable clothing and footwear, and make the case again at the general meeting held each Sunday evening in Pompeii. Although most pharmaceuticals are available in Italy, many medicines require prescriptions and opening hours of local pharmacies often do not coincide with our free time. We, therefore, keep a limited quantity of first aid supplies and recommend that team members bring all prescription medicines with them as well as any other medicines that they feel they may need. The hotel staff have been a great help in assisting with health problems that have arisen. During the 2008 season when two of our team members required medical assistance, the hotel staff arranged for a doctor to visit them and later for hospital visits as required. While many team members are experienced travelers, recent years have seen increasing numbers of younger individuals who may not have the same range of experience.

11.4   Postfieldwork

Before departing Italy, we submit to the Soprintendenza Archaeologica di Pompei a DVD containing a summary of our work for that season and a copy of our database in order to keep him informed and up-to-date about our work. This is not only a courtesy, but a necessary requirement to work in Pompeii. In addition, the Project Director prepares and distributes to Staff and team members of the previous season a Field Report covering the accomplishments of the preceding summer and places a summary on the website. And, finally, the responsible staff member updates the Project website with the dates for the following summer. Upon completion of these three essential responsibilities, planning begins for the new season.

11.5   Summary of Lessons Learned

The Pompeii Food and Drink Project focuses on a clearly defined research plan, but recognizes the importance of providing an educational experience for those who participate. PRs and Staff have learned much about organizing, planning, and conducting an archaeological surface investigation in a foreign country with many individuals of wide-ranging ages, from different backgrounds, and with varying levels of expertise and experience in archaeology. Many important challenges we anticipated; others we did not and had to overcome through experience by trial and error. Among the most critical lessons learned are the following:

1. Working in a foreign country with intelligent people of different ages from different backgrounds that range from student to educated layman to skilled professional requires that PRs and Staff exercise patience and respect.
2. Early, detailed, and thoughtful planning is necessary for the efficient and productive functioning of the Project. Include as many Staff personnel as possible in decision making, so that they feel part of the process. A key problem we have labored to resolve is planning the workday to ensure that team members have enough work to occupy their time and of a quality to engage their interest and talents.
3. Communication at all levels and among all participants is critical to accomplish the goals of the Project, to maintain high morale, and to ensure a successful and enjoyable season. Of particular importance is keeping PRs and Staff informed throughout the process from the planning stage, through the actual work, and between seasons.
4. Flexibility is an absolute necessity. A well-planned project must allow for the unexpected in organization, work schedule, technology, and individual personality. The Italian work schedule and labor activity, for example, differ markedly from what is normal in the United States, so due allowance for unanticipated delays or cancellations of long-planned activities must be made.

Conducting an archaeological field school in Pompeii these past 10 years has presented many challenges. But, with proper planning, diligent oversight, and professional conduct, the experience has provided abundant rewards to all participants.

Acknowledgments For authorization to study, for the provision of permits, and for daily assistance on-site we especially thank Dottoressa Teresa Elena Cinquantaquattro, Superintendent, and Dottore Antonio Varone, Director of Excavations at Pompeii; Dottoressa Grete Stefani, Director of Excavations and Antiquarium at Boscoreale; and the staff and custodians, especially Signor A Boccia, of the Soprintendenza Speciale per i Beni Archeologici di Napoli e Pompei. We also thank Dottoressa Mariarosaria Salvatore (Former Superintendent); Dottore Pietro Giovanni Guzzo (Former Superintendent); and Dottore Antonio D'Ambrosio (Former Director) for their assistance in past years. Dottoressa Eugenia Salza Prina Ricotti graciously has offered us encouragement and advice from the beginning of the project and provided us access to her personal notes on kitchens in Pompeii.

Reference

Pompeii Food and Drink Project web site http://www.pompeii-food-and-drink.org/. Retrieved April 14, 2011.

Chapter 12
Historical Archaeology Artifact Training in Field Schools: Three International Case Studies

Alasdair Brooks

12.1 Introduction

The present author has worked on four different continents at different stages of his career and has been involved with historical archaeology field schools within three; North America, Europe, and Australia. The following chapter offers a summary of the author's experiences teaching artifact processing and analysis in historical archaeology (his primary area of specialization) in each of these field schools: Thomas Jefferson's Poplar Forest, Virginia, USA (1994–1996); Castell Henllys, Wales, UK (1998–2001); and Port Arthur, Tasmania, Australia (2003–2004).

The chapter is not intended to be a comprehensive comparative study of the teaching of artifact work within field schools in the three nations covered here. The three field schools offer only a very small selection of relevant educational opportunities. While Port Arthur has traditionally been the only recurring annual historical archaeology field school in Australia, Poplar Forest and Castell Henllys are both only one of many field schools in their respective countries. Furthermore, the author only worked at the three institutions for relatively brief periods of the field schools' history; the experiences discussed here may not be relevant for the periods before and after the author's period of involvement. This study is therefore by necessity selective and – based as it is on the author's personal experiences – somewhat subjective.

Despite these caveats, discussion here stems from the premise that a comparison of how artifact analysis has been taught at these three institutions reflects important differences regarding the role of, and attitudes towards, historical artifact analysis generally within their respective countries. The United States, United Kingdom, and Australia have all developed very different approaches to the analysis of artifacts

A. Brooks (✉)
School of Archaeology and Ancient History, University of Leicester,
University Road, Leicester LE1 7RH, UK
e-mail: amb71@le.ac.uk

H. Mytum (ed.), *Global Perspectives on Archaeological Field Schools:*
Constructions of Knowledge and Experience, DOI 10.1007/978-1-4614-0433-0_12,
© Springer Science+Business Media, LLC 2012

dating from the last 500 years ("historical archaeology" being used here in the American and Australian sense: the archaeology of the modern world after 1500), and these broader themes inevitably inform how artifacts are approached at both the professional and educational levels, and to what extent they are emphasized in vocational training.

12.2 The Three Sites

12.2.1 Poplar Forest

Thomas Jefferson's Poplar Forest is a historic house museum just to the southwest of Lynchburg, in Bedford County, Virginia. As the name suggests, the property once belonged to Thomas Jefferson, third president of the United States. Jefferson inherited the property from his father-in-law John Wayles in 1773 (Chambers 1993: 4), and in 1806, built the octagonal retreat home that now forms the center of the modern property (Chambers 1993: 35–36). Jefferson's grandson Francis Eppes inherited the property in 1826, though he moved to Florida just 2 years later (Chambers 1993: 167, 175). After serving as a private residence for the subsequent 160 years, the core of Jefferson's once much-larger property was acquired in 1983 by the current owners, the Corporation for Thomas Jefferson's Poplar Forest (Chambers 1993: 208).

The Corporation began an ambitious program of restoration on Jefferson's octagonal retreat and the landscape. While it had been in constant use as a family home for 160 years, the appearance of the building had been changed following a fire in 1845 (Chambers 1993: 181–183); many of the original outbuildings had also altered or were missing, with only the two matching octagonal privies remaining significantly unchanged since Jefferson's day. Important transformations had also occurred to the property's inner landscape, which Jefferson had carefully designed according to what were then the latest ideas of symmetry in European landscape design (Brown 1990). Today, the house and adjacent utilitarian "wing of offices" appear much as they did in Jefferson's day, and ongoing archaeological work on the landscape continues to uncover information about the ornamental grounds; the latest information on this research can be found on the Poplar Forest website (http://www.poplarforest.org).

From almost the very beginning, archaeology formed an important part of the restoration, initially with support from the archaeology program at Jefferson's somewhat better-known Monticello property (today an hour and a half's drive, but several days' travel in the early nineteenth century), but soon established as an independent archaeology program at the property. Much of the research initially focused on the areas immediately around Jefferson's house, helping to uncover the wing of offices (Chambers 1993: 82–83) and undertaking important work about the central plantation landscape. While this work remained important, in the mid-1990s the focus of much of the archaeology program was on the plantation's enslaved community, and the archaeology field schools of this period spent much of their time working on the "Quarter Site" (Heath 1999; Heath and Bennett 2000), consisting of four late eighteenth-century slave-associated structures predating the construction of Jefferson's retreat home (though not Jefferson's ownership of the property). The field school – which had been running for 22 years as of the summer of 2010 – has been accredited by the University of Virginia throughout its existence.

12.2.2   Castell Henllys

Castell Henllys is an Iron Age and Romano-British hill fort located in the Pembrokeshire Coast National Park between Fishguard and Cardigan in Wales. Several of the Iron Age roundhouses have been reconstructed, and the hill fort is open to the public, managed via the National Park. Archaeology has been an important part of Castell Henllys since the 1980s, and from 1985 to 2008, the University of York provided an annual field school, alongside a volunteer program, as part of the long-running summer excavation and research season (see also Chap. 7).

The University of York's field school and volunteer programs were international, with students from the United States, the United Kingdom, continental Europe, and even Australia. The North American and British students were offered slightly different programs, with the North American students offered a course structured around the requirements of the North American university credit system, and the British students offered opportunities in keeping with the requirements of their courses.

At first glance, an Iron Age hillfort in Wales may not seem like the most likely location for a field school in historical archaeology, and indeed the archaeology of the modern world was never the primary focus of the Castell Henllys training excavations. But from the 1980s, excavation director Harold Mytum ran a historical archaeology research program alongside the main excavations (see Chap. 7). This partially consisted of work on historic gravestone commemoration (Mytum 1994, 1999), but also included excavations at several nineteenth-century cottages undertaken in the late 1980s (Mytum 1988) – the ceramics from which were the subject of research by the current author (Brooks 2002, 2003) – and, from the mid 1990s, fieldwork at Henllys Farm (Fig. 12.1), a nearby Tudor through nineteenth-century farm complex (Mytum 2010). Artifact analysis for these often occurred on-site, and where possible was integrated into teaching and training opportunities for both North American and British students. While historical archaeology was neither the main research focus nor the main teaching focus of the Castell Henllys field school, it was therefore an important additional aspect of the ongoing program during the author's involvement.

12.2.3   Port Arthur

Port Arthur, located in the southeast of the Australian island state of Tasmania, is one of the most famous, some might say infamous, historical sites in all of Australia.

Fig. 12.1 Ceramics being excavated at Henllys Farm, Pembrokeshire, by a student on the Castell Henllys Field School (photograph copyright Harold Mytum)

Tasmania was settled (or invaded) by the British in 1803. Over the next half a century, nearly 70,000 convict men, women, and children were transported to the new colony (Jackman 2009: 101). Port Arthur was specifically founded as a penal station for recidivist convicts in September 1830. By the time the site was closed in 1877 – 24 years after convict transportation to Tasmania had ended – approximately 10–12,000 convict sentences had been served at the site, with a peak convict population of 1,128 reached in 1846; indeed, the Tasman Peninsula where Port Arthur is located was the third most populous district in the colony in the 1840s (Jackman 2009: 102; Tuffin 2004: 77). Though it was never a typical convict site, Port Arthur remains a highly emotive and iconic site in Australia, where the convict past is often a difficult and contested issue. Some of the broader themes surrounding these divergent views of the site, and how this has sometimes impacted archaeological interpretation, have been explored in detail by Jackman (2009). As early as 1915, the former convict station was becoming a popular tourist destination; this despite damage from bush fires, and attempts to turn the site into a residential town. The state government began to reacquire the land, and by 1940, all of the surviving convict buildings were back in government hands. The Australian National Parks and Wildlife Service began a comprehensive program of site improvements in the 1970s, and the Port Arthur Historic Site Management Authority (or PAHSMA), the current site management authority, was formed in 1987 (Jackman 2009: 102).

Archaeology has long been an important part of site management, conservation, and interpretation. The University of Sydney undertook an important test program in the 1970s, and in the 1980s – thanks to the Port Arthur Conservation and Development Project that immediately preceded PAHSMA's management – Port Arthur became recognized both nationally and internationally as an important archaeological site (Egloff 1986). The site was also vitally important in the development of historical archaeology in Australia. While now replaced by more recent research, the original *Port Arthur Conservation and Development Project Archaeological Procedures Manual* (Davies and Buckley 1987) long served as the standard benchmark reference for historical archaeology best practice in the country. Indeed, so influential was Port Arthur for Australian historical archaeology management that Ireland (2004: 102) referred to the site as the "conscience" of national heritage management.

Archaeology enjoys a high public profile during the summer, and for several years this has included an annual summer field school for Australian university students, the only recurring single-site historical archaeology field school in the country. Many Australian students are required to gain a minimum number of hours of practical experience in order to study for an honors degree. A standard Australian undergraduate degree is for 3 years; a limited number of exceptional students then go on to study for an honors degree in an additional fourth year (see Chap. 5). This experience can sometimes be hard to find in Australia given the relatively limited number of excavation opportunities, particularly in historical archaeology. While a majority of the Port Arthur students are from Victoria – just across Bass Strait from Tasmania – during the author's professional interaction with the site there were also strong contingents from South Australia and Queensland. South Australia's Flinders University, which has an internationally known maritime archaeology program, has at times also run a separate maritime archaeology field school at Port Arthur, but the current discussion focuses solely on the land-based program.

12.3 Teaching Artifacts at the Three Field Schools

12.3.1 Poplar Forest

During the period of the author's involvement, the Poplar Forest field school lasted 5 weeks in late June and early July; the timing typically allowing for the integration of archaeology into American independence celebrations on the Fourth of July. During this time, the primary focus of the students' work was on fieldwork, not artifact processing or analysis. However, the students were rotated (typically in pairs) through the on-site archaeology lab, where they were given hands-on experience in the processing (initial cleaning and labeling) of artifacts under the supervision of the lab supervisor. While the cataloging was all undertaken by the lab supervisor, ad hoc training was also provided for basic training artifact identification during the processing. These periods in the archaeology lab were supplemented by specialist lectures from Poplar Forest archaeology staff and site visits to other archaeology labs during fieldtrips to other sites in Virginia. Further opportunities for both hands-on experience and teaching were provided by periods of rain. These weather-related opportunities for further artifact work are a shared experience at all three field schools (and are no doubt near-universal in nonarid climates), though at Poplar Forest afternoon thunderstorms were the most common cause of precipitation.

While the artifact and lab component at Poplar Forest may not have been the primary focus of the students, the artifact training component was carefully integrated into both the field school and the broader archaeology program. The site archaeology program had (and still has) a fulltime on-site archaeology lab supervisor who was responsible for artifact processing, cataloging, and analysis. In addition, the archaeology lab was located in a converted historic (though not Jefferson-era) barn, one side of which featured a large window which was used for exhibits on artifact analysis, and which allowed the public to see the artifact analysis being undertaken by the archaeologists (though cleaning and labeling was, in the author's time, typically undertaken in an alcove out of public view).

The high visibility afforded to artifact work in public interpretation was also transferred to the field school teaching materials. Each field school student was given a copy of the in-house archaeology manual, updated annually; in the mid-1990s, this updating was typically undertaken by the lab supervisor and director of archaeology. Significant sections of this manual were given to artifact identification and dating, including seven pages devoted to Poplar Forest's basic in-house artifact processing procedures (Andrews et al. 1996: 13–19) and another 20 pages given over to simple typologies and dates for ceramics, glass, and metal (Andrews et al. 1996: 20–39). Combined, these 27 pages were three times longer than the nine pages devoted to field procedures and site etiquette (Andrews et al. 1996: 13–19). The typologies were in turn supported by bibliographies and were based on common terminologies drawing on well-known work from Noël Hume (1970), Miller (1991), Jones and Sullivan (1989), and other iconic foundational material culture research that would be familiar to most archaeologists working in the Mid-Atlantic region of the United States. This manual therefore offered an important supplement to hands-on teaching of artifact work and helped to emphasize the extent to which material culture work is an important component of historical archaeology, even if the bulk of students' time in a 5 week field school was necessarily directed towards learning basic field techniques.

12.3.2   Castell Henllys

Historical archaeology artifact identification and analysis was not as fully integrated into the Castell Henllys field school as it was at Poplar Forest. This simply reflects the extent to which historical archaeology was itself just one option for field schools students who could select either the historic, late prehistoric, or Roman period site for their main training location; it was also an option, or supplement to the primary Iron Age and Romano-British site, for the UK training excavation participants. To

Fig. 12.2 Training by the author in glass identification, using an assemblage from a Pembrokeshire cottage site (photograph copyright Harold Mytum)

the extent that the topic was taught at Castell Henllys in the late 1990s, it often depended on the availability of a University of York research student interested in the subject, and then hired on an ad hoc basis for the summer, rather than the presence of a fulltime permanent member of staff. It was not a core focus of the teaching materials and reading given to students, though neither was it ignored by these materials.

The Castell Henllys field season typically lasted for 6 weeks, divided into two groups of 3 weeks, though participants could choose to stay for the full period (and the North American students were required to do so). Artifact processing and teaching usually took place in a marquee erected next to the main campsite (Fig. 12.2), which was shared by a historical archaeology-oriented research student and a final-year undergraduate supervising the processing of the finds from the hill fort. This marquee was usually large enough to accommodate 20 or so students, and so also served as an excellent ad hoc lecture theater. Most of the students who worked in the "artifact tent" did so on a volunteer basis, though southwest Wales has a considerably damper climate than central Virginia, so there were also more opportunities to gain experience during rainy days at Castell Henllys. This experience was supplemented by occasional evening lectures, at least one of which was typically devoted to historical archaeology and material culture.

The main exceptions to this pattern were the North American field school students, who were provided with additional structured lectures on historical archaeology artifacts, which included hands-on training on seventeenth through nineteenth-century ceramics using a teaching collection generated by the present author from the artifacts excavated at Henllys Farm. The students would also typically spend 2 weeks engaged in gravestone recording in Ireland, gaining further experience in material culture research; this gravestone research was a central component of the field school, reflecting one of the primary research interests of the director (Mytum 1994, 1999). This research, however, was organized separately from the work in Wales, and the present author had no involvement in this aspect of the field season. Some artifact training was also provided on glass when suitable collections of the latter were available. The North American students were also required to write an extended research essay during the field school, and many chose to focus on historical archaeology artifact subjects.

In a sense, the historical archaeology artifact training at Castell Henllys very much depended on the continent of origin of the individual student. Historical archaeology may not have been the core focus of the excavation, but within what was a wholly appropriate limitation given the nature of the broader site, North American students were intentionally introduced to – and given some experience in – the same type of artifacts and artifact terminologies that they would have been exposed to had they chosen to study in the United States or Canada. The other students could – if they chose to – go through the entire 6 weeks of the training excavation with almost no exposure to post-1500 material culture except for the occasional evening lecture and rain day; nonetheless, the opportunity was there for them to gain more training in this type of material culture if they were interested. Again, historical archaeology was not necessarily the focus of the training excavation for UK students; it was an additional opportunity attached to the main excavation. For many British and European students with other research interests in the Eastern Atlantic past, there would have been little perceived need for more extensive exposure to either historical archaeology or historical archaeology artifact analysis.

12.3.3   Port Arthur

As is the case with so many other aspects of Australian archaeology (and indeed broader cultural life), the teaching of artifact analysis at Port Arthur in the mid-2000s reflected aspects of both the North American and the British experiences, but with distinctively Australian elements. As with Poplar Forest, historical archaeology was the sole focus of the field school; Australian archaeology, like North American archaeology, is largely divided between the historical archaeology of European settlement and the archaeology of the pre-European indigenous population. As with Castell Henllys, there was no permanent artifact specialist or lab supervisor on staff; the field school lab supervisor was hired for the summer on a temporary short-term contract. The two fulltime archaeologists on staff during the period of the author's involvement were both field archaeologists. This reflected a general shortage of material culture specialists in Australia rather than an intentional decision to sideline artifact analysis. Indeed, Greg Jackman, the site's head at the time, undertook extensive work to improve and more fully integrate material culture analysis into the site's archaeology. The present author first became involved with artifact analysis at Port Arthur when he was asked to help update the site's artifact database structure and terminology alongside Australian archaeologist Catherine Tucker. Perhaps the most significant difference between Port Arthur and the other two field schools under discussion here is that the Tasmanian institution had no formal university link in the mid-2000s. There were nonetheless many informal links, and many, perhaps most, of the students were from La Trobe University in Melbourne, Victoria, with significant minorities coming from the University of Queensland in Brisbane, and Flinders University in Adelaide, South Australia.

As with Poplar Forest, each student at the field school was required to work in the on-site archaeology lab on a rotational basis, gaining experience in the cleaning and labeling of finds. While cataloging was again undertaken only by the lab supervisor (who was typically hired to work for a couple of additional weeks either side of the field school), the students did gain experience in how to identify and classify the artifacts they were working with. At Port Arthur, this consisted exclusively of nineteenth and twentieth-century artifacts, as – shipwrecks aside – no Euro-Australian sites outside Sydney and Norfolk Island (both founded in 1788) predate the nineteenth century. As with the other sites, additional artifact processing experience was generally available when it rained.

While students were provided with a reading list, there was no field school teaching manual at Port Arthur. By the first decade of the twenty-first century, the once-iconic Port Arthur archaeology manual (Davies and Buckley 1987) had not been updated in over a decade. Several sections were now seriously out of date, and this was particularly true of the artifact identification and classification discussions, which failed to reflect the exciting developments in Australian artifact analysis that had been underway since the late 1990s. At the time of the author's involvement with the field school, efforts were underway to completely rewrite the manual alongside the total redevelopment of the site artifact database. The present author was therefore involved with Port Arthur at what appeared to be a transitional time. If artifact processing and analysis during the field school were not necessarily a primary focus of the student training, nor as integrated into broader site interpretation as at Poplar Forest, important work was underway to update and improve this aspect of the archaeology program, both for the field school specifically and the site generally.

12.4   Field Schools and Broader Artifact Analysis in North America, Britain, and Australia

The differences between the approach of the three field schools to teaching artifact processing analysis reflects that the role of artifact analysis has developed very differently within historical (or, in Britain, postmedieval) archaeology in these three countries. This is perhaps equally true of historical archaeology generally, but there are issues specific to artifact analysis that lie at the core of how the field schools approached artifacts within their educational environments.

As a recognized subdiscipline with a locally organized professional society, historical archaeology is about the same age in each country. The North America-based Society for Historical Archaeology and the UK-based Society for Post-Medieval Archaeology were both founded in 1967 (Barton 1967; Pilling 1967); the Australasian Society for Historical Archaeology (originally the Australian Society) followed just 5 years later in 1972. Originally, the path of post-1500 artifact analysis followed similar paths in the Atlantic world. Description and categorization of the material culture encountered by the emerging subdiscipline was paramount, and it is perhaps significant here that one of the most prominent (and still in print) early guides to artifact analysis – Noel Hume's *Guide to the Artifacts of Colonial America* (1970) – was written by a British archaeologist working in the United States.

The most extensive comparative historiography of artifact analysis in the UK, North America, and Australia (Brooks 2005: 3–14) notes that the arrival of the New Archaeology, and in particular the publication of South's *Method and Theory in Historical Archaeology* in 1977, marks a real break in British and North American approaches to artifact analysis – and, by association, artifact teaching. Another significant difference stemmed from the well-known disciplinary differences; North American archaeology was considered a subdiscipline of anthropology, while British archaeology has always remained an entirely independent discipline (though one with strong interdisciplinary links), and South was not alone in developing a distinctive North American approach. But *Method and Theory* arguably marks the point where the developing differences in approaches to artifacts were crystallized in print. The book is well known for arguing strongly for an approach based on formulating testable hypotheses based on data collation; as South noted in a recent interview (Joseph 2010), artifacts were to be considered crucial as the base data underlying this hypothesizing paradigm. This approach would soon combine with the equally important interpretive paradigm developed near-simultaneously by Deetz, most prominently in *In Small Things Forgotten* (originally published in 1977), and North American historical archaeology would come to consider artifacts as important not just for typologies and chronologies, but as an important tool in formulating site interpretation. This naturally transfers to the teaching of artifact analysis within field schools; even where teaching fieldwork techniques are central to a historical archaeology field school, the integration of at least some basics of artifact identification is considered vitally important given the importance of artifacts as underlying interpretive data.

British postmedieval archaeology would, at least until the second half of the 1990s, and the publication of *The Familiar Past* (Tarlow and West 1999) continue to focus on typologies, chronologies, and issues of production as the central aspect of artifact work (Brooks 2005: 3–4); a few prominent examples of the latter might include Coleman-Smith and Pearson's (1988) work on the Donyatt kilns, Barker's (1991) work on the Staffordshire potter William Greatbach, and Gaimster's (1997) monumental description of German stoneware from 1200 to 1900. This descriptive work was itself vitally important, as the more interpretive work favored by North Americans would arguably be impossible without a solid and well-researched descriptive base. More recently, however, North American and British approaches to artifacts have arguably been moving closer together. Interpretive studies of artifacts, or larger projects incorporating such approaches, have recently become far more common (Brooks 1999; Casella 2009; Wilson 2008), to the extent, indeed, that there have even been theoretical critiques of the "interpretive paradigm" within British historical archaeology (Jeffries et al. 2009: 328–329). The traditional dichotomies between anthropology/archaeology and interpretation/description that have stereotypically framed discussion of North American and British historical archaeology are arguably no longer as important as before.

From the perspective of teaching post-1500 artifacts, the more significant difference is that postmedieval archaeology is simply not as central to British and European archaeology as it is within North American archaeology. Where North American archaeology has traditionally been divided between prehistoric archaeology and historical archaeology, European archaeology is divided along a much larger number of periods. Prehistorians, Romanists, medieval archaeologists, and postmedievalists all have a role. These periods are often further subdivided; many medievalists, for example, will specialize in either the early Saxon period or the later periods following the Norman conquest of 1066. Within this broad spectrum of period-based specializations, the post-1500 period has often been peripheral. Unlike in the United States, where prominent historical archaeologists are often based within contract archaeology – note, for example, work by Charles Cheek (Cheek and Friedlander 1990) and Seifert (1991) both of the Northern Virginia office of John Milner Associates – it is very unusual to find a British contract archaeologist who specializes almost exclusively in the post-1500 period. Furthermore, when a contract archaeology firm is excavating a multiperiod site, focus has traditionally very much been on the earlier periods. While there are increasingly important exceptions, the limited available funds were typically directed towards the prehistoric, Roman, and Medieval contexts. Within this environment, the role of both historical archaeology generally and artifact analysis specifically at the Castell Henllys field school can be seen as symbolic of the broader British whole. Students were exposed to both and could gain experience in both if they were interested (Fig. 12.3), but it would arguably unnecessarily limit the career prospects of an 18-year-old undergraduate student to only teach them the archaeology of a period that is often peripheral to the professional environment outside of academia.

Australia has its own distinct history of artifact analysis, though Australian historical archaeology shares characteristics with both North America and Britain. Like in North America, archaeology is usually divided between prehistoric (or indigenous) archaeology and historical archaeology, though the important research field of the archaeology of indigenous missions (Ash et al. 2008; Lydon 2009; Lydon and Ash 2010) arguably bridges the gap between the two with more relative frequency than similar studies in North America. As in Britain, archaeology is an independent discipline rather than a subdiscipline of anthropology. Yet unlike both of the latter, artifacts were simply not a central component of Australian historical archaeology

Fig. 12.3 British students washing historic period artifacts at Castell Henllys (photograph copyright Harold Mytum)

for much of its development, something which was occasionally a point of concern (Birmingham 1988: 149; Brooks 2005: 1–2; Lawrence 1998). While there were important early studies on trade (Allen 1978) and descriptions of local pottery traditions (Ioannou 1987), for almost a decade between 1987 through to 1998, there were very few papers on material culture in the journal *Australasian Historical Archaeology* (Brooks 2005: 8). Connah's (1988) guide to Australian historical archaeology, "*Of the Hut I Builded*"; *The Archaeology of Australia's History*, reflects this broader state of affairs by containing almost no discussion of artifacts. Instead, industry – particularly mining – was often a primary focus of Australian research, a research focus which was certainly not inappropriate given the centrality of the post-1850 Gold Rush to Australia's colonial development. There was little explicit engagement with either British or American approaches to artifact work in this period, with the notable exception of a paper by Bavin, which explicitly drew on work by Deetz and Deagan in a study of status and class in Australian material culture (Bavin 1989: 16).

A 1998 paper by Susan Lawrence titled "The Role of Material Culture in Australasian Archaeology" played a crucial role in arguing that artifact analysis could, and should, play a more central role in Australian historical archaeology. Since that year, the description and analysis of artifacts has become far more important to the regional subdiscipline, with important publications on local and international ceramics (Casey 1999; Brooks 2005; Hayes 2007), glass (Carney 1998; Davies 2006), and holistic site studies incorporating artifact analysis into broader site interpretation (Brooks et al. in press; Casey 2004; Lawrence 2000) all playing an increasingly important role. Yet it should be stressed that, compared to North America and Britain, this is a relatively new phenomenon – something reflected in the publication date of the citations in the last sentence. Also unlike North America and Britain, there are almost no dedicated material culture specialists in Australia. This is largely a matter of simply demographics. Australia is a country the size of the continental United States with a population a third of that of the United Kingdom, and with far fewer professional archaeologists than either of the latter countries. Given the geographic scope of potential work, and the relatively small number of archaeologists available to carry it out, most professionals in the country do not have the luxury to become artifact (or indeed other types of) specialist; they are professional necessity generalists. In the mid-2000s, the Port Arthur field school epitomized this broader professional situation. Artifact analysis was seen as increasingly important to the teaching program, and solid proactive steps were being taken to improve this aspect of the field school experience. At the same time, artifact work was less integrated into the overall field school environment than was the case at Poplar Forest, and training was very much focused on preparing students for a professional environment where generalists were naturally favored over specialists.

These broad differences in approaches to material culture work are also mirrored by differences in field technique. This is not the place to discuss these in detail, but they are potentially important to this discussion. In the United States, screening of excavated soils is extensive. A long tradition of recognizing the importance of horizontal spatial distributions in plowed soil (Brooks et al. 2009: 40) means that North American historical archaeologists regularly go so far as to screen agriculturally plowed soil. While artifacts are recorded in a combination of grid reference and context, a glance at any of the standard guides to British field techniques, such as Roskams' *Excavation* (2001), will quickly demonstrate that British field archaeology takes a contrasting approach. Screening is comparatively rare – unusual, even – while priority is given to recording to context, with grid references rarely used. Australia, again, is somewhere in between the two, with one recent study of plowzone archaeology in Australian historical archaeology explicitly drawing on plowzone techniques more commonly associated with North America, and geophysical testing and surface mapping techniques more commonly associated with Britain and Europe (Brooks et al. 2009). But there is no set "national" approach to screening soil for artifacts on Australian sites; some excavations use a more American approach, others use a more British approach. These differences in field technique may also impact how artifacts are studied; it might be argued, for example, that the American approach facilitates the type of extensive spatial quantification often favored in American archaeology via its rigid classification by grid. Certainly, these differences by necessity impact approaches to processing; teaching students a context-first recording approach in Britain might well prove confusing if they go on to work in a grid-first recording environment in North America, while teaching artifact analysis in Australia arguably requires familiarity with both approaches.

12.5   Conclusion

The three very different structures of these field schools in three very different parts of the world therefore help to form, and are informed by, the three very different general approaches to historical archaeology and artifact analysis in each of the home countries. Perhaps this is self-evident as a general observation, but the specifics for each individual country remain illuminating. Broadly speaking, the Poplar Forest field school's approach to artifacts reflected the established status of material culture within a historical archaeology that is arguably better established within both academic and popular culture than its British or Australian counterparts. The Castell Henllys approach reflected both the established status of specialist artifact analysis within overall British archaeology, but the arguably somewhat more marginal role of post-medieval studies within the broader spectrum of British periodized archaeology; it also reflected the more international nature of the participating students (Fig. 12.4). The Port Arthur approach reflected the general lack of material culture specialists within an Australian historical archaeology that was only beginning to come to grips with the enormous potential of material culture studies within a subdiscipline that had traditionally been forced by both demographics and geography to train generalists rather than specialists.

Fig. 12.4 Cataloging finds as part of the Castell Henllys Field School (photograph copyright Harold Mytum)

Field schools are very much part of their broader disciplinary environment; very few are able to transcend the methodological, disciplinary, and theoretical backgrounds of their host institution's home region. But perhaps to some extent this is as it should be. While field schools certainly can take a role in shaping broader disciplinary debates, surely their primary duty – other than their duty to the archaeological record – is to give their students basic training in how to carry out archaeology in their home country. Poplar Forest students gained a strong disciplinary background in the approaches relevant to the Mid-Atlantic United States, learning how artifact analysis is traditionally integrated into the region's historical archaeology. Castell Henllys students were made aware of the potential of postmedieval artifact studies and were given the opportunity to study the topic more closely if it interested them, but within a multiperiod environment appropriate to British archaeology. Port Arthur students were shown the potential of the developing field of Australian colonial material culture studies, but within the context of preparing for a vocational environment where generalists are needed over specialists. The observation that the three field schools are products of their specific environments does not therefore necessarily imply a value judgment that one approach was better than another; they all succeeded in their duty to their students within that broader disciplinary environment.

References

Allen, J. (1978). The archaeology of nineteenth-century British imperialism: An Australian case study. In R. Schuyler (Ed.), *Historical archaeology: A guide to substantive and theoretical contributions* (pp. 139–148). Farmingdale: Baywood Publishing.

Andrews, S. T., Brooks, A., Fischer, L., Heath, B., Patten, D., & Strutt, M. (1996). *Poplar forest archaeology lab and field manual 1996. Manuscript on file*. Forest, VA: The Corporation for Thomas Jefferson's Poplar Forest.

Ash, J., Brooks, A., David, B., & McNiven, I. (2008). Catalogue of European manufactured objects found at the "Early Mission Site" of Totalai, Mua Island. *Memoirs of the Queensland Museum, 4*(2), 473–492.

Barker, D. (1991). *William Greatbach, a Staffordshire potter*. London: Antique Collector's Club.

Barton, K. (1967). The origins of the society for historical archaeology. *Post-Medieval Archaeology, 1*, 102–103.

Bavin, L. (1989). Beyond the Façade: The expression of status and class in material culture. *The Australian Journal of Historical Archaeology, 7*, 16–22.

Birmingham, J. (1988). The refuse of empire: International perspectives on Urban Rubbish. In J. Birmingham, D. Bairstow, & A. Wilson (Eds.), *Archaeology and colonisation: Australia in the world context*. Sydney: Australian Society for Historical Archaeology.

Brooks, A., Lawrence, S., & Lennon, J. (in press). The parsonage of the Reverend Willoughby Bean: church, state and frontier settlement in 19th-century colonial Australia. *Historical Archaeology 45*(4).

Brooks, A. (1999). Building Jerusalem: Transfer-printed finewares and the creation of British identity. In S. Tarlow & S. West (Eds.), *The familiar past? Archaeologies of later historical Britain* (pp. 51–65). London: Routledge.

Brooks, A. (2002). The cloud of unknowing: Towards an international comparative analysis of eighteenth- and nineteenth-century ceramics. *Australasian Historical Archaeology, 20*, 48–57.

Brooks, A. (2003). Crossing Offa's Dyke: British ideologies and late eighteenth- and nineteenth-century ceramics in Wales. In S. Lawrence (Ed.), *Archaeologies of the British; Explorations of identity in Great Britain and its colonies 1600–1945* (pp. 119–137). London: Routledge.

Brooks, A. (2005). *An archaeological guide to British ceramics in Australia, 1788–1901*. La Trobe University Archaeology Program, Melbourne, and the Australasian Society for Historical Archaeology, Sydney.

Brooks, A., Bader, H.-D., Lawrence, S., & Lennon, J. (2009). Ploughzone archaeology on an Australian historic site: A case study from South Gippsland, Victoria. *Australian Archaeology, 68*, 37–44.

Brown, C. A. (1990). Thomas Jefferson's poplar forest: The Mathematics of an ideal villa. *Journal of Garden History, 10*(2), 117–139.

Carney, M. (1998). A cordial factory at Parramatta, New South Wales. *Australasian Historical Archaeology, 16*, 80–93.

Casella, E. C. (2009). "You knew where you were": An archaeology of working class households in turn-of-century Cheshire. In A. Horning & M. Palmer (Eds.), *Crossing paths or sharing tracks; Future directions in the archaeological study of post-1550 Britain and Ireland* (pp. 365–380). Woodbridge: Boydell.

Casey, M. (1999). Local pottery and dairying at the DMR Site, Brickfields, Sydney, New South Wales. *Australasian Historical Archaeology, 17*, 3–37.

Casey, M. (2004). Falling through the cracks: Method and practice at the CSR site, Pyrmont. *Australasian Historical Archaeology, 22*, 27–43.

Chambers, S. A., Jr. (1993). *Poplar forest and Thomas Jefferson*. Forest, VA: The Corporation for Thomas Jefferson's Poplar Forest.

Cheek, C., & Friedlander, A. (1990). Pottery and pig's feet: Space, ethnicity and neighborhood in Washington DC 1880–1940. *Historical Archaeology, 23*(1), 34–60.

Coleman-Smith, R., & Pearson, T. (1988). *Excavations in the Donyatt Pottery*. Chichester: Phillimore.

Connah, G. (1988). *"Of the Hut I Builded"; The archaeology of Australia's history*. Cambridge: Cambridge University Press.

Davies, P. (2006). Mapping commodities at Casselden Place, Melbourne. *International Journal of Historical Archaeology, 10*(4), 336–348.

Davies, M., & Buckley, K. (1987). *Port Arthur conservation and development project archaeological procedures manual*. Occasional Paper No. 13, Department of Lands, Parks, and Wildlife Tasmania, Hobart.

Deetz, J. (1977). *In small things forgotten: The archaeology of early American life*. New York: Anchor Books.

Egloff, B. (1986). *The Port Arthur story: 1979 to 1986 (being a true and accurate account in brief of the port Arthur conservation and development project)*. Hobart: National Parks and Wildlife Service.

Gaimster, D. (1997). *German stoneware 1200–1900*. London: British Museum Press.

Hayes, S. (2007). Consumer practice at Viewbank homestead. *Australasian Historical Archaeology 25*, 87–104.

Heath, B. J. (1999). *Hidden lives; The archaeology of slave life at Thomas Jefferson's poplar forest*. Charlottesville: University of Virginia Press.

Heath, B. J., & Bennett, A. (2000). "The Little Spots Allow'd Them": The archaeological study of African-American yards. *Historical Archaeology, 34*(2), 38–55.

Ioannou, N. (1987). A German Potter in the Barossa Valley, South Australia. *The Australian Journal of Historical Archaeology, 5*, 29–40.

Ireland, T. (2004). The Burra Charter and historical archaeology: Reflections on the legacy of Port Arthur. *Historic Environment, 18*(1), 25–28.

Jackman, G. (2009). From stain to saint: Ancestry, archaeology, and agendas in Tasmania's convict heritage – a view from Port Arthur. *Historical Archaeology, 43*(3), 113–118.

Jeffries, N., Owens, A., Hicks, D., Featherby, R., & Wehner, K. (2009). Rematerialising metropolitan histories? People, places and things in modern London. In A. Horning & M. Palmer (Eds.),

Crossing paths or sharing tracks; Future directions in the archaeological study of post-1550 Britain and Ireland (pp. 323–349). Woodbridge: Boydell.

Jones, O., & Sullivan, C. (1989). *The Parks Canada glass glossary*. Ottawa: Parks Canada.

Joseph, J. W. (2010). An interview with Stanley A. South. *Historical Archaeology, 44*(2), 132–144.

Lawrence, S. (1998). The role of material culture in Australasian archaeology. *Australasian Historical Archaeology, 16*, 8–15.

Lawrence, S. (2000). *Dolly's creek; An archaeology of a Victorian goldfields community*. Melbourne: Melbourne University Press.

Lydon, J. (2009). Imagining the Moravian mission: Space and surveillance at the former Ebenezer Mission, Victoria, Southeastern Australia. *Historical Archaeology, 43*(3), 5–19.

Lydon, J., & Ash, J. (2010). The archaeology of missions in Australasia: Introduction. *International Journal of Historical Archaeology, 14*(1), 1–14.

Miller, G. (1991). A revised set of CC index values for classification and economic scaling of English ceramics from 1787–1880. *Historical Archaeology, 25*(1), 1–25.

Mytum, H. (1988). The Clydach Valley, a nineteenth-century landscape. *Archaeology Today, 9*(3), 33–37.

Mytum, H. (1994). Language as symbol in Churchyard monuments: The use of Welsh in nineteenth- and twentieth-century Pembrokeshire. *World Archaeology, 26*, 252–267.

Mytum, H. (1999). Welsh cultural identity in nineteenth-century Pembrokeshire: The Pedimented headstone as a graveyard monument. In S. Tarlow & S. West (Eds.), *The familiar past? Archaeologies of later historical Britain* (pp. 215–230). London: Routledge.

Mytum, H. (2010). Biographies of projects, people and places: Archaeologists and William and Martha Harries at Henllys Farm, Pembrokeshire. *Postmedieval Archaeology, 44*(2), 294–319.

Noël Hume, I. (1970). *A guide to artifacts of colonial America*. New York: Alfred A. Knopf.

Pilling, A. (1967). Beginnings. *Historical Archaeology, 1*, 1–12.

Roskams, S. (2001). *Excavation*. Cambridge: Cambridge University Press.

Seifert, D. (1991). Within sight of the White House: The archaeology of working women. *Historical Archaeology, 25*(4), 82–108.

South, S. (1977). *Method and theory in historical archaeology*. New York: Academic Press.

Tarlow, S., & West, S. (Eds.). (1999). *The familiar past? Archaeologies of later historical Britain*. London: Routledge.

Tuffin, R. (2004). Cascades probation station: Prison built on timber. *Tasmanian Historical Research Association, 51*(2), 70–83.

Wilson, R. (2008). "The mystical character of commodities": The consumer society in 18th- century England. *Post-Medieval Archaeology, 42*(1), 144–156.

Part V
Fieldwork and People

Chapter 13
From Graduate to Professor: Changing Perspectives on Field Schools

Bonnie J. Clark

The need for a book on field schools was never so apparent to me as when I began research for this chapter. There are many versions of the "archaeology field handbook," some of them classic (Barker 1977), some tongue-in-cheek (Praetzellis 2003), and others encyclopedic (Burke, et al. 2009). *Collaborating at the Trowel's Edge* focuses on the role of the field school in indigenous archaeology (Silliman 2008), exploring the relationship between pedagogy and indigenous populations, something with which we should all be concerned. The book is especially useful for those interested in field schools that involve collaboration with any community, indigenous or not. Jennifer Perry's longitudinal study at the San Clemente field school insightfully looks at the role of the field school in archaeological training and practice (2004). It contains some helpful hints about running a field school and should be required graduate school reading. But given the crucial role that field schools play in the discipline, and particularly their importance in the professional life of graduate students and professors, the limited range of both good practical advice and epistemological exploration of field schools in general emerges as a disciplinary black hole. Until recently, no such resources existed for the archaeology field school, though Baxter (2009) now provides some important frameworks for planning and implementing field school teaching. However, the variability of aims, practice, context, and traditions means that much can be learnt from a wide variety of experiences, as indicated throughout this volume. This chapter is a reflexive study about how experience in implementing a field school is gained and applied across the critical divide between graduate school and professorate.

The analysis presented here is largely derived from my own experience, which is as a student and practitioner of Americanist anthropological archaeology. However, my view is broadened by both informal and more formal correspondence with other archaeologists who have recently crossed this same divide. The more formal discussion

B.J. Clark (✉)
Department of Anthropology, University of Denver, Denver, CO, USA
e-mail: bclark@du.edu

H. Mytum (ed.), *Global Perspectives on Archaeological Field Schools:*   217
Constructions of Knowledge and Experience, DOI 10.1007/978-1-4614-0433-0_13,
© Springer Science+Business Media, LLC 2012

involved a snowball questionnaire sent to tenure-track professors who either are on their way to tenure or who have recent achieved it (all described here as "junior" faculty). The questions posed of them were:

- What is the relationship between your dissertation research and (a) the first field school you taught? (b) the most recent field school you taught?
- What was the most important prior experience or training for running your own field school? What training or experience do you wish you would have had?
- How critical do you think teaching a field school is to your overall success as an academic (e.g. Was it a requirement for being hired? Was it/will it be important for tenure?).
- Do you have any anecdotes to share or advice for those preparing to teach their first field school?

What emerges from the answers to these questions, and my own experience, are three themes. First, the training that graduate students receive in implementing a field school is inconsistent at best. The relationship between dissertation research and field schools is illustrative in this sense. This situation is particularly problematic because of the second theme: the ways in which field schools play a key role in early professional life of academics. That role is both instrumental, meaning it is critical for gaining jobs and tenure, and also cultural, it is at the heart of who academic archaeologists are. Finally, short of a substantial overhaul of graduate student training, there are some relatively simple interventions that could make a significant difference to the lives of practitioners. Those easily implemented modifications need to be accompanied, however, with more difficult disciplinary discussions about the relationship between academic archaeologists and the field school.

13.1 A Second Rite of Passage?

For our undergraduates, the field school is seen as a rite of passage, a "definitive moment" (Perry 2004:236) during which they either decide to be archaeologists or to pursue cleaner, more lucrative professions. Running your own field project could be considered a second rite of passage, where the apprentice becomes the master. In reality that transition is often not so clear cut, and that is as it should be. When asked about their training for running their own field school, many junior faculty identify their time as teaching assistants on academic projects or as crew chiefs doing cultural resource management (CRM). Many of us had some more middle-ground experience, serving as assistant or field project directors, especially for projects for which we had earlier had served in another, less senior capacity. The case study presented by Perry (2004) shows how that transition from student to staff member can work within a long-term field school of over ten seasons (see also Chaps. 1, 2, 6, 8, and 14).

The extent to which prior supervisory experience was useful when running a field school for the first time depended significantly on the level of involvement and commitment to pedagogy by those in charge of previous fieldwork experience. Apprentice field staff need a lot of help and supervision themselves; this is a node of overlap between the quality of the undergraduate and graduate field experience (Perry 2004:254). The more that senior staff are invested in actively mentoring students, the better. Transparency is also an important issue; letting both undergraduate and graduate students see and even participate in the decision-making processes not only increases engagement, it also helps apprentice archaeologists make similar decisions in the future (see Chap. 9). For example, almost anyone who has spent time in the field has had the experience of excavating empty test units. This apparent failure is instead a great opportunity to explain sampling methodologies and the importance of negative data, while also letting students know that their work, which is seemingly unproductive, is just the opposite.

It is important that many junior faculty do not, however, point to experiences as crew chiefs as their most important preparation. As many of us who direct projects know the hard way, logistics and personnel issues can make or break a field school (Baxter 2009). It may well be that time spent as a camp counselor or in the Boy Scouts might be better preparation in the long run than acting repeatedly as an assistant with some field responsibilities. One junior faculty who pointed to the Boy Scouts as his best training for running a field school wrote "Really … it fits with archaeology, which draws so much experience and technology from the military." For my part, growing up in a big family taught me how to manage cramped living quarters with little privacy, how to get big groups to and from places in a relatively timely manner, and how to shop and cook for a crowd.

The relationships between dissertation research and field school experience reflect the very diverse experience that is graduate field training. Some Ph.D. students can shape a field school around their own research using their institutional affiliation to recruit students; others craft an independent field project, but use volunteers rather than students (see Chaps. 1, 2, 6, and 14). My own dissertation research while I was at the University of California, Berkeley, was hybrid of the two. The first field season included students taking the class for field school credit, as well as volunteers recruited through the University Research Expedition Program, a volunteer program similar to Earthwatch but affiliated with all of the University of California campuses. My field work was largely funded by fees paid by both the students and the volunteers (Fig. 13.1). Because outside grant funding was secured the next year, my second year of field work included only one student taking the class for credit, while the rest of the workforce were returning volunteers from the previous year and a few fellow graduate students, creating a more diverse and more experienced crew.

There can be significant pitfalls to allowing graduate students to craft and implement their own field projects. This is a situation that works best when proper support structures are in place to avoid failure, such as guaranteed resources or field oversight by faculty or other mentors. As one of my respondents wrote, "It also helped me develop my field strategies and decision-making abilities as a graduate student to have my advisor 'on call' if I needed advice or a site visit, but not to have that individual present at all times to supervise me." The dissertation field school should not be a sink-or-swim situation because the stakes are too high for the students, the

Fig. 13.1 Author with field school volunteers and students on site in southeastern Colorado, Summer 2000

archaeological resources, and the institution. Even seasoned field workers can be overwhelmed by adding the pressure to recover good, relevant data on top of the usual field stressors. As I once wrote of my first dissertation field season, "Maybe by the time someone has led a number of projects they really believe they are going out to test hypotheses; the first time out it's clear your hypotheses will be testing you" (Clark 2001).

Many archaeology Ph.D. students choose a dissertation topic related to a long-standing field project of which they have already been a part. Often these are run as field schools, but not always; field schools are not welcome everywhere. For example, Mexico does not allow international field schools, in part because officials do not want their cultural patrimony used to make money for outside institutions. This means students who conduct dissertation research in Mexico might obtain supervisory experience there, but not the experience of either participating in or staffing a field school. This is also true for students who choose topics that are either not field-based or are not amenable to group study. This may limit their fieldwork options later in their careers, though this may be compensated by other experiences that provide opportunities to develop successful careers.

Very few archaeologists go on to teach their own field school at their former dissertation site once they are junior faculty, especially for those whose dissertation research was part of a larger project run by an institution to which they now do not have an affiliation, but it is also true for those whose work was more independent. In some cases, this is clearly driven by administrative desires for faculty to develop field schools either nearby or affiliated with their new home institution. In other cases, it is linked to changing research priorities or logistical issues that make return difficult if not impossible.

13.2   The Field School Lynch Pin

Unfortunately, there has only been a small sample of Assistant Professor of Archaeology positions advertised recently. Of those listed on the Society for American Archaeology website in the Fall of 2010, just under half of them specifically mentioned active field research as a qualification. Of those requiring a field project, the majority indicated that involving students in field research was an important criterion for the position. Although it may not be explicitly spelled out in the job description, there is often an expectation that junior faculty will teach a field school. A colleague who has held two junior faculty positions was asked about field schools during both of his interviews; he remarks "I can't imagine if I'd said anything but 'yes, I love field school' I would have gotten anywhere with the committee." In contrast, other junior faculty have had the opposite experience, where they had to fight to teach field courses. Why senior faculty and administrators discourage the practice varies. Certainly, field work is both time- and energy- intensive, both resources that are at a premium for all junior faculty. Because tenure deliberations focus so strongly on individual contribution, the inherently collaborative nature of most fieldwork can be seen as a liability. One could also cynically point to the devaluing of the field school as symptomatic of the overall underrepresentation of teaching in the typical tenure portfolio.

Despite the disparity of institutional positions about field schools, junior faculty still identify them as key to their academic success and happiness. As one of my respondents wrote, "Teaching the field school is one of the best parts of my job." In fact, one could argue it is often a keystone professional activity. For many junior faculty, a successful field school significantly raises their campus visibility. Field schools involve both active student participation and applied work. Both of those factors make field schools attractive to university administrators and public relations staff; photographs of smiling field school students often grace campus literature. This is particularly the case for junior faculty who have been able to craft a field school that involves a local or regional community. As one colleague puts it, "Field schools always attract press attention and locals love to hear that archaeologists are interested in their history."

The very collaborative quality of fieldwork that some administrators see as a drawback can ironically be very positive when it comes to establishing junior faculty professionally. Working together in the field creates camaraderie among students, and it does the same for faculty. The networks forged in consultations, in co-taught courses, and in searching for collaborators on grants serve junior faculty well when they come up for tenure. That applies whether the connections are within a department, across a campus, or between institutions.

Field schools often become the engine that drives a faculty member's long-term research program and that appears to be as true for those teaching at schools without a graduate program as those with one. This seems somewhat counterintuitive, given the role future faculty themselves played as graduate student teaching assistants, but sometimes making way for graduate student research through field schools can detract from junior faculty priorities. If the new academic's research has to be somehow shoehorned in around an unrelated field school, it tends to take a backseat at the very time that it is most needed. This is where most junior faculty need shrewd judgment to find a new field site at which to work as a teacher and where they can recover data that link to their own research goals. The path to their own research becomes just that much more difficult because they now need to invest significant time and effort in scouting a new field locale and planning an appropriate research design.

In the generally enthusiastic embrace of the field school by junior faculty, there are some dark clouds hovering over where field schools fit into the nexus of institutional expectations. This is an area where students could use more mentoring regarding the academic job search and subsequent position negotiations. Any department that expects a junior faculty member to teach field schools needs to be willing to support them, whether that is through start-up packages, through course release during the regular year, or through collaboration with more established faculty. If the expected course is unrelated to faculty research, there needs to be compensatory time allowed to create space for personal research in order that they can maintain their overall career development. Even with field schools that support junior faculty research goals, an annual summer field school can quickly lead to burnout. Junior faculty themselves might willingly commit to this arrangement in the beginning, but they would be well advised not to.

13.3   Plugging the Gaps

In his article on archaeological pedagogy, Yannis Hamilakis points out what he sees as a great paradox, "the disjuncture between the highly developed and lively theoretical debates on archaeological research, on interpretive, ontological, epistemological and political aspects on archaeology and the poverty of theorizing with regard to teaching and learning" (2004:293). It is worth asking ourselves just what a field school is and how do we help prepare our students to craft and implement them.

We intuitively know something of an answer to the first question. The field school is a locale of excursionary learning in archaeological field practice. It is very often also a primary method by which research archaeology is conducted. As Perry points out, the finite time and budget restraints of a field school can create tensions between the needs for student education and data recovery (2004). For the Register for Professional Archaeologists to grant certification to a program, "the primary objective of an academic field school must be the training of students" (Register for

Professional Archaeologists 2003). However, when someone's thesis, dissertation, tenure case, or promises to funding agencies are on the line, the balance can tip in the other direction. In these situations, many junior faculty wish they had received better and more explicit training on how to juggle those priorities and stay within a budget. It is hard to imagine taking graduate students in the field and not providing them a copy of the research design. But how often do we do that and not provide them with a copy of the budget? Just like transparency around other field logistics, transparency about budgetary priorities would be a step in the right direction.

Another issue comes to the surface when junior faculty are asked where their training was lacking: safety. One suspects that some junior faculty learned about the importance of field safety the hard way. Their advice to first time field school directors often mentioned safety issues including, "No dig is ever worth risking the safety of your students, workers, or volunteers or their health." Another added, "Take every injury seriously, however minor it may seem." This is one area where academics could take a few cues from CRM firms. Many CRM firms require that anyone in a supervisory position in the field maintain up-to-date first aid and workplace safety training. While it would seem that campus risk management would see the value in this, getting this same training for graduate students and faculty often means rattling a few administrative cages, particularly if funds end up coming out of already stretched budgets. While some guidance can be provided on health and safety issues for field schools in general (Baxter 2009), it is necessary that formal (preferably certified) training is provided for key field staff and that risk assessments are drawn up and distributed for any project to protect all involved.

The ability to manage interpersonal relationships in the field school is another key issue raised by the survey. One respondent wished that she had received some training in counseling, while another pointed to difficulty, especially for new field directors, of finding the balance between discipline and fun. Perry's surveys of field school participants (2004) indicate personality conflicts are one of the most common problems of a field school and she pointedly writes "Losing the enthusiasm and motivation of the students can be potentially ruinous" (2004:254). Yet on reflection, field school participants often see the trying conditions that can lead to personnel clashes – long work hours, close quarters, working closely in teams – as opportunities for personal growth. For my part, I am clear with my field school participants that they are embarking on participant observation in small group living. Carefully screening future participants and not being afraid to tackle problems early are both suggestions that came out of my survey. Another suggestion comes from Perry, who identifies periodic changes in crew composition and rotation of responsibilities and leadership as other ways to check tensions around the power structures of field hierarchy (2004:253). Many of the same issues that arise in field school are not dissimilar to those that the staff of student housing face on a regular basis. Most campuses have people on hand who are skilled in counseling students and in conflict mediation. We should take better advantage of these potential sources for training and support and should also make sure that interpersonal skills and strategies are a strong element of our mentoring for field school staff.

Improved training in the three key areas of budgeting, safety, and personnel management would not only better prepare graduate students to take the lead roles in their own field schools, it would also make them better prepared to work in CRM. These are among the basic management skills that the Society for American Archaeology's Task Force on Curriculum identified as important to graduate education (Bender and Smith 2000), but have rarely been formally addressed. Fieldwork competence, of which this management is a part, has also been identified as a key concern of archaeological educators in Australia (Colley 2004). There has been considerable angst that most graduate programs are geared toward academics, but most practitioners work in CRM (see e.g., Fagan 2000). This is one of the many places of shared territory between the academy and CRM; all our students need these skills, regardless of their professional destination.

13.4 Confronting the Field School

Identifying strategies to plug some of the gaps in how we train our graduate students is important. But as Hamilakis (2004) points out, an instrumental focus on pedagogy must also be accompanied by a deeper questioning of disciplinary epistemology. Although the traditional foci of field schools has been on teaching and research, a third is now emerging: the field school as a way for scholars and institutions to meet community needs (Nicholas 2008). In this way, the field school is part of the paradigm shift toward community engagement and service learning taking place on campuses across the U.S. (Campus Compact 2010) as well as a groundswell within our field. In her 2008 Distinguished Lecture to the Archaeology Division of the AAA, Alison Wylie made the claim that collaborative archaeology is the new, "new archaeology;" in other words, it is the archaeology of the future (Wylie 2008). The authors in the Silliman volume (2008) see the field school as a central locale for this new type of practice, but they are not alone (see, e.g. Little and Shackel 2007; Nassaney and Levine 2009; Shackel and Chambers 2004; and many contributors to this volume).

A commitment to community engagement comes through strongly in my interactions with junior faculty, with many seeing the connection to those communities involved in their work as critically important. In some ways, we can look to this generation of junior faculty as vanguards of this paradigm shift. While many of them were trained in field schools with some community participation, their own field schools have made this a central concern. One of my respondents suggests that transparency about community involvement is as important as transparency about more traditional research design. Many junior faculty work to bring members of different communities together as field crew, what Silliman calls "collaborating at the trowel's edge" (2008). Others are thinking broadly about the diverse ways they interact with many communities, not just descendant communities, but also local individuals, from landowners to shopkeepers. By making this a key concern, junior faculty are positioning their own students to be leaders in this new wave as well.

The level of collaboration and community engagement in an archaeological field school is not the only area where disciplinary soul searching should be taking place. There are many good reasons to embrace the field school, as evidenced by the experiences of junior faculty. But we should be careful of the overvalorization of field work at the expense of other disciplinary practices, including laboratory analysis, work with museum collections, archival research, and writing (see Chap. 4). As Robert Preucel points out, the association of archaeology with fieldwork, especially excavation, is so strong that the two are essentially synonymous (Joyce and Preucel 2002:19). As such, it might play an overly important role in the identity of junior faculty, who spend much more time as teachers than as archaeologists. This may be one reason why they often push themselves to the limit to run a field school. It is what attracted them to archaeology as undergraduates, and later on in their careers it reminds them they are archaeologists and not just professors. So while junior faculty are often pushing the boundaries of collaboration, they might not be questioning the centrality of field practice. All of us should be considering whether our field schools can and should also be located in laboratories, archives, or museums (see Chap. 10).

Incorporating the other end of archaeology – the museum where our objects end up – is very much a practical and ethical necessity for the field school that I developed as a junior faculty member. The field school is part of a long-term community-engaged research project at the Granada Relocation Center, one of ten Japanese American internment camps build in the western U.S. during World War II. Better known as Amache, the site is owned by the adjacent town of Granada but maintained by the Amache Preservation Society (APS), a group of high school students led by their social studies teacher. The APS holds a significant collection of historic documents and artifacts, some of which were donated by former internees, some from staff of the camp, and some that derive from earlier archaeological research at the camp. The APS has done wonderful community outreach using these collections, but when I first got involved there was no comprehensive system in place for managing them. The creation of such a system was identified in a preliminary development plan as one of the most pressing issues faced by site managers (Ellis 2004). Archaeologists have an ethical obligation to ensure that the artifacts we collect are property curated (Lynott and Wylie 2000). Helping the APS organize their collections and set up collections management systems not only met a previously identified need, it also meant that the collections from the field school would also be protected (Fig. 13.2).

The field school has been held twice at Amache and the APS museum. Each year, we have spent the mornings in the field and the afternoons at the museum. As an element of an archaeological field school, incorporating the museum has been remarkably successful. The project is able to draw in some students primarily interested in museum work along with those who were looking for training in historical archaeology and community-engaged scholarship. Students receive valuable training in the management of a small museum including collections management, object handling and storage, label and exhibit text writing, and engaging with museum visitors. In addition, the students' appreciation of the materials we recover

Fig. 13.2 Field school students and teaching assistant working in Amache museum, Granada, Colorado, Summer 2008

each morning in the field is greatly enhanced by spending their afternoon engaging with complete objects and associated historic documents in the museum.

The Amache experience suggests incorporating the museum into archaeological training is not just ethically sound, it is also good pedagogy. In this suggestion I am certainly not alone. Many archaeologists have embraced the site museum as an element of field schools (e.g. Joyce 2003; Chap. 9). This is particularly true of those interested in community collaboration (e.g. Hantman 2004). Museums are critical locations of public archaeology and our research lives longer there than almost anywhere else (Moyer 2007). They are also valuable community resources, often helping turn archaeological sites into sustainable heritage sites.

13.5 Final Thoughts

This volume on field schools should instigate a larger disciplinary discussion about the role of field schools in research, in pedagogy, in epistemology, and in the culture of anthropology. The topic covered by this chapter, how graduate students and new professors engage with the field school, certainly deserves more study and more reflexive publications. One of my colleagues surveyed for this article suggested that he would have loved a "playbook" when he ran his first field school. There is an interesting resonance here with my experience mentoring graduate student teaching assistants this past summer. Although I had created a handbook for the students of the field school, my graduate students pointed out that I did not make one for them as field supervisors. It turned out that they could have benefited just as much, if not more so, from their own written playbook. So I include myself in the challenge when I say that as a discipline we need to be more conscientious about what our graduate students learn from and about field schools.

While the graduate guide is not yet written, my colleagues have provided some good advice to apprentice and yeoman directors of field schools. It is with their suggestions that I end this chapter:

- Be flexible and be prepared.
- Incorporate community interaction.
- Have fun and be real.
- Make anthropology relevant.
- Know when to be a jerk.
- Be clear about the rules of conduct.
- Plan, but prepare to be surprised.

and perhaps the most important,

- Don't pull through the drive through liquor store with university insignia on the doors or license plates.

Acknowledgments I would like to thank the junior faculty who have been willing to share their experiences with me, whether that has been at a conference, over a frosty beverage, or via the internet. My more senior colleagues and mentors have likewise been an incredible resource. We should not underestimate the worth of that collective wisdom. A hearty thanks to my own field crews through the years, without whom none of this learning would have happened.

References

Barker, P. A. (1977). *Techniques of archaeological excavation.* New York, NY: Universe Books.
Baxter, J. E. (2009). *Archaeological field schools. A guide for teaching in the field.* Walnut Creek, CA: Left Coast.
Bender, S. J., & Smith, G. S. (Eds.). (2000). *Teaching archaeology in the twenty-first century.* Washington, D.C: Society for American Archaeology.
Burke, H., Smith, C., & Zimmerman, L. J. (2009). *The archaeologist's field handbook.* Lanham, MD: AltaMira.
Campus Compact. (2010). Who we are. (Electronic Document). Retrieved October 15, 2010, from http://www.compact.org/about/history-mission-vision/
Clark, B. (2001). *Epistemology and fear: Constructing the first field project.* Paper presented at the Society for Historical Archaeology annual meeting, Long Beach, CA.
Colley, S. (2004). University-based archaeological teaching and learning and professionalism in Australia. *World Archaeology, 36*(2), 189–202.

Ellis, S.M. (2004). *Camp Amache, Prowers County, Colorado: Site management, preservation, and interpretive plan*. SWCA Environmental Consultants, Salt Lake City, UT. Submitted to the Denver Optimists' Club.

Fagan, B. M. (2000). Strategies for change in teaching and learning. In S. J. Bender & G. S. Smith (Eds.), *Teaching archaeology in the twenty-first century* (pp. 125–131). Washington, D.C.: Society for American Archaeology.

Hamilakis, Y. (2004). Archaeology and the politics of pedagogy. *World Archaeology, 36*(2), 287–309.

Hantman, J. L. (2004). Monacan meditation: Regional and individual archaeologies in the contemporary politics of Indian heritage. In P. A. Shackel & E. J. Chambers (Eds.), *Places in mind: Public archaeology as applied anthropology*. New York, NY: Routledge.

Joyce, R. A. (2003). Working in museums as an archaeological anthropologist. In S. D. Gillespie & D. L. Nichols (Eds.), *Archaeology as anthropology* (Monograph 11). Arlington, VA: Archaeology Division of the American Anthropological Association.

Joyce, R. A., & Preucel, R. W. (2002). Writing the field of archaeology. In R. A. Joyce (Ed.), *The languages of archaeology* (pp. 18–38). Oxford: Blackwell.

Little, B. J., & Shackel, P. A. (Eds.). (2007). *Archaeology as a tool of civic engagement*. Lanham, MD: AltaMira.

Lynott, M. J., & Wylie, A. (Eds.). (2000). *Ethics in American archaeology* (2 Revised ed.). Washington, D.C.: Society for American Archaeology.

Moyer, T. S. (2007). Learning through visitors: Exhibits as a tool for encouraging civic engagement through archaeology. In B. J. Little & P. A. Shackel (Eds.), *Archaeology as a tool of civic engagement* (pp. 263–277). Lanham, MD: AltaMira.

Nassaney, M. S., & Levine, M. A. (Eds.). (2009). *Archaeology and service learning*. Gainesville, FL: University Press of Florida.

Nicholas, G. P. (2008). Melding science and community values: Indigenous archaeology programs and the negotiation of cultural difference. In S. W. Silliman (Ed.), *Collaboration at the trowel's edge: Teaching and learning in indigenous archaeology* (Amerind Studies in Archaeology, pp. 228–249). Tucson, AZ: University of Arizona Press.

Perry, J. E. (2004). Authentic learning in field schools: Preparing future members of the archaeological community. *World Archaeology, 36*(2), 236–260.

Praetzellis, A. (2003). *Dug to death: A tale of archaeological method and mayhem*. Walnut Creek, CA: AltaMira.

Register for Professional Archaeologists. (2003). Guidelines and standards for archaeological field schools. (Electronic document). Retrieved October 13, 2010, from http://www.rpanet.org/associations/8360/files/field_school_guidelines.pdf.

Shackel, P. A., & Chambers, E. J. (Eds.). (2004). *Places in mind: Public archaeology as applied anthropology*. New York, NY: Routledge.

Silliman, S. W. (Ed.). (2008). *Collaborating at the trowel's edge: Teaching and learning in indigenous archaeology*. Tucson, AZ: University of Arizona Press.

Wylie, A. (2008). *Legacies of collaboration: Transformative criticism in archaeology: The archaeology division distinguished lecture*. Presented at the annual meeting of the American Anthropological Association, San Francisco, CA.

Chapter 14
Suvoyuki Means Joint Effort: Archaeologists, the Hopi Tribe, and the Public at Homol'ovi

Lisa C. Young

14.1 Introduction

Archaeologists in the United States increasingly view information sharing and collaboration with local and descendent communities as critical components of field projects (Colwell-Chanthaphonh and Ferguson 2008; LaBelle 2003; Singleton and Orser 2003). Archaeological field schools are an ideal setting for students to gain experience in public outreach and to participate in collaborative activities. Despite this, however, only a small proportion of the field training programs offered in the United States during the first decade of the twenty-first century included these types of opportunities. In this chapter, the rewards and challenges of including public outreach and collaboration with descendant communities in the field school experience are discussed. My case study is the Homol'ovi area of northeastern Arizona (Fig. 14.1) where archaeologists and Hopi community members have interacted with each other on archaeological projects for over a century. A recent partnership between the Hopi Tribe and Arizona State Parks, and the development of the Homol'ovi Undergraduate Research Opportunities Program (HUROP), has encouraged *suvoyuki* – a Hopi word meaning "joint effort."

14.2 Archaeologists, Hopi People, and the Public at Homol'ovi

Until the passage of the Native American Graves Protection and Repatriation Act (NAGPRA), archaeologists in many areas of the United States had limited contact with the descendent Native communities whose ancestral remains they studied. In the American Southwest, however, Indian reservations are a prominent part of the

L.C. Young (✉)
University of Michigan, Museum of Anthropology, Ann Arbor, MI, USA
e-mail: lcyoung@umich.edu

H. Mytum (ed.), *Global Perspectives on Archaeological Field Schools:*
Constructions of Knowledge and Experience, DOI 10.1007/978-1-4614-0433-0_14,
© Springer Science+Business Media, LLC 2012

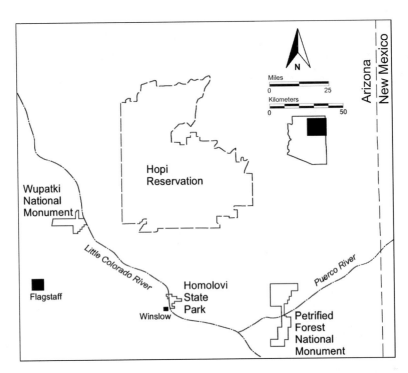

Fig. 14.1 Location of Homol'ovi State Park and Hopi Reservation in Northeastern Arizona

cultural landscape. Many tribes reside on at least part of their traditional homelands, and archaeologists have a long history of interacting with tribal community members. Archaeologists interested in the history of ancestral puebloan people have used oral traditions and cultural practices of contemporary tribes, such as the Hopi, Zuni, Acoma, Laguna, and the numerous pueblos in the Rio Grande River area of New Mexico, to inform their interpretation of archaeological remains for over a century. Tribes in the Southwest also have a long history of interacting with the public through the sale of crafts and tourism that began with the establishment of the transcontinental railway across Arizona and New Mexico in the 1890s (Howard and Pardue 1996).

In the Homol'ovi area, the relationship between Hopi people and archaeologists began in the late nineteenth century and changed during the twentieth century as new challenges and opportunities arose. Jessie Walter Fewkes (1898) conducted the first excavations at Homol'ovi. Fewkes learned about the sites in the area from Hopi clan migration stories. During Fewkes' research and collecting expeditions for the Bureau of American Ethnology, Hopi informants from First and Second Mesa villages told him that their ancestors had lived in large villages along the Little Colorado River, roughly 65 miles (105 km) to the south, before migrating to their present-day villages. Hopi people continue to call this area Homol'ovi, which means place (*ovi*) of the small hills (*homol*) and describes the small sandstone buttes that are so common in and around the town of Winslow, Arizona. Fewkes asked Hopi men from these clans to show him where the ruins of their ancestral villages were located and then hired them as workers on his excavation projects. Even though their ancestors left these sites roughly five centuries before, the Hopi workmen recognized the layout of the villages from the ruins, which in turn helped Fewkes decide where to excavate.

Following Fewkes' fieldwork, little professional research was conducted in the Homol'ovi area, but local interest in the sites grew. One of the sites is located just north of the city of Winslow and is known to locals as "Pottery Hill." Throughout the midtwentieth century, pot hunting occurred at all the large pueblo sites at Homol'ovi and in other surrounding areas. Heavy equipment was used at Homol'ovi II, the largest of the Homol'ovi pueblos with over 1,000 rooms (Walker 2000). Over the years, many individuals expressed concern about the destruction of these sites. By the early 1980s, Bruce Babbitt, the governor of Arizona at the time, formed an advisory board of Hopi people, archaeologists, and Winslow residents to make recommendations about the best way to protect these sites. The board recommended the creation of a state park, which was approved by the Arizona State legislature in 1986. The Homol'ovi Ruins State Park[1] opened to the public in 1993.

In conjunction with the formation of the State Park, funds were also allocated to the Arizona State Museum at the University of Arizona for research, management, and interpretation of cultural resources in the Park. Dr. E. Charles Adams was hired to direct these efforts, and the Homol'ovi Research Project (HRP) was created. Adams' goals included public involvement in archaeological research and developing relationships with the Hopi Tribe (Adams 2000:2). Since its inception, volunteers have been included in HRP field programs and laboratory analyses. The formation of the State Park also provided opportunities for the general public to visit archaeological excavations and a chance to interact with archaeologists. When working on sites easily accessible to Park visitors, archaeologists gave tours of their excavations on a daily basis and during an annual open house, called "Archaeology Day."

Starting in the mid-1980s, archaeological field research at Homol'ovi also provided opportunities for archaeologists to interact with members of the Hopi Tribe. During the early years of the project, HRP provided summer internships for several high school students from Hopi, who worked on the excavations and archaeological survey of the Park. Hopi community members also visited the excavations during the field season. These visits provided an opportunity for archaeologists and Hopi people to share their perspective on these ancestral villages with each other, and Hopi community members often offered prayers and cornmeal at the conclusion of the field season. It was in this context that I, as a graduate student supervising excavation units, became aware of just how powerful the connection between Hopi people and these archaeological sites was and still is. Hopi people view archaeological sites as footprints of their ancestors and a living connection to their accomplishments and journeys.

---

[1] Archaeologists spell Homol'ovi with an apostrophe, following the spelling in the Hopi Dictionary (Sekaquaptewa et al. 1998). The State Park name does not use the apostrophe. In 2011, Arizona State Parks decided to drop the word "ruins" was dropped from the Park's name.

During the 1990s, visits by community members became less frequent. There are many reasons for this change; elders passed on, and HRP field seasons shortened, often ending during a very busy part of the Hopi ceremonial cycle. As a result, it difficult for Hopi people to find time to visit the excavations. Beginning with the development of the Hopi Cultural Preservation Office (CPO) in 1988 (Ferguson et al. 2000) and following the passage of the NAGPRA in 1990, relations between archaeologists and the Hopi Tribe also became more formalized and focused on consultation. In the early 2000s, the Hopi Tribe expressed its desire to have closer connections with the Homol'ovi State Park, which the Tribe viewed as a gateway to the Hopi Mesas. In 2003, the Hopi Tribe signed a memorandum of understanding with Arizona State Parks creating a partnership to further develop and interpret the resources at the Homol'ovi State Park through increased Hopi community involvement. In 2006, the Hopi Tribe hired a project manager to be the liaison with State Parks and began a more active role in planning events at Homol'ovi, such as the annual open house. The Tribe also created a summer internship at Homol'ovi for Hopi college students.

As this brief summary shows, Homol'ovi has been a place where archaeologists, Hopi people, and the public have interacted for over 25 years. Recent initiatives by the Hopi Tribe to play a more active role in the management of the Park have created new opportunities to expand and reinvigorate these relationships. As I developed the Homol'ovi Undergraduate Research Opportunities Program, I built on these opportunities and the long history of interactions.

14.3 Public Outreach and Collaboration in a Field School Setting

The goal of the HUROP was to provide undergraduate students with hands-on experience in archaeological field research along with real world training in public outreach. HUROP also introduced students to Hopi culture and traditions through interactions with tribal members and visits to the Hopi Mesas. In addition to myself, the program also included five faculty mentors who were archaeologists or museum studies professionals and a team of graduate students who were excavation crew chiefs, laboratory supervisors, or museum studies interns. Besides learning about archaeology by participating in excavations, students also attended evening lectures and went on weekend field trips to learn about the past in northeastern Arizona from archaeological and Native community perspectives, as well as the challenges of interpretation. Incorporating public outreach opportunities and interactions with Hopi tribal members into the field school setting created a dynamic and rich learning environment for students, which enhanced their archaeological field training experience.

Field school directors elsewhere have integrated Native community concerns and perspectives in their research objectives and curriculum (see chapters in Silliman 2008). These programs are often labeled "Indigenous Archaeology" (Silliman 2008:2) and are explicitly designed to help meet the needs of a tribe, often focusing on heritage management (Chilton 2006; Mills et al. 2008) or political status issues (Rossen 2008; Silliman and Dring 2008). In developing HUROP, I chose instead to focus on public outreach and interpretation as a venue for collaboration with the Hopi Tribe. Although my research questions focused on understanding the organization of ancient communities, this emphasis also made the research accessible to Native communities and the general public. Each year my research designs were reviewed during consultations with the Hopi Tribe and seven other tribes with connections to the Homol'ovi area.

The inclusion and integration of archaeological and indigenous perspectives in a field school curriculum must be predicated on mutual respect for different ways of knowing about the past. Through evening talks and discussions, Hopi archaeologists and individuals with knowledge of clan histories introduced HUROP students to Hopi oral history and perspectives on archaeology. Hopi people, including several youth groups, also came to visit our excavations. On site tours, students were responsible for presenting information on the areas where they were excavating. These contexts provided a forum for visitors to ask questions and for the students to share what they were learning.

One of the HUROP faculty mentors, Susan Secakuku, who is a museum professional and a member of the Hopi Tribe, was vital to the collaborative and educational efforts of the project. Shortly after becoming a faculty mentor for HUROP, Susan was hired by the Hopi Tribe as the Homol'ovi Park Project director. Susan became the liaison between Arizona State Parks and the Tribe who oversaw a partnership that enabled further development and interpretation at Homol'ovi. The Hopi Tribe also wanted to increase Hopi community involvement at Homol'ovi. With these goals in mind, Susan and I worked with Arizona State Parks staff to develop public outreach activities that would jointly meet the goals of the Hopi Tribe and offer opportunities to HUROP students.

Although HUROP students participated in a wide variety public outreach opportunities during three field seasons (2006–2008), I focus my discussion on projects that highlight the rewards and challenges of intertwining interpretation and archaeological field training into a field school curriculum. The first example examines changes to the Park's annual open house and highlights broader shifts in the relationship between the archaeologists who work at Homol'ovi and Hopi people. The second example is the construction of a traditional Hopi corn roasting pit, a project which created opportunities for HUROP students to work directly with Hopi community members and to learn how living traditions can be used to interpret archaeological remains. My final example explores how the creation of an exhibit case in the Park visitor center laid the foundation for future collaborative interpretations.

14.3.1 Changes to the Annual Open House

The initial motivation to develop an open house at the Homol'ovi State Park was to provide opportunities for the public to see the excavations and research undertaken

Fig. 14.2 HUROP student talks to *Suvoyuki* Day visitor about artifacts from the excavations

by archaeologists at Homol'ovi. The original name of this open house was "Archaeology Day." Over the years, the open house expanded to include not only site tours and talks by archaeologists, but also activities that allowed visitors to learn about Hopi culture, such as a pottery firing, an early morning run, and corn roasted in a pit. With the development of the Homol'ovi Park Project, the Hopi Tribe requested that the name of this event be changed to "*Suvoyuki* Day." *Suvoyuki* means "joint effort" and was chosen to emphasize the new partnership between the Hopi Tribe and Arizona State Parks. The event was also expanded to include booths with a variety of Hopi artists, nonprofit organizations from the Hopi Mesas, and traditional foods. Performances by a local Hopi dance group that included youth were also added to the schedule. *Suvoyuki* Day attracted hundreds of visitors to the Park, including tourists, local Winslow residents, Hopi people, archaeologists, and anthropologists.

At *Suvoyuki* Day, HUROP students were able to share what they had learned about archaeology at Homol'ovi. Near the artists' and informational booths, we set up a table where the students washed artifacts from our excavations and talked to visitors. I asked the students to help visitors understand what archaeologists can learn from various types of artifacts and to discuss the importance of archaeological context with the underlying message that looting destroys information critical to interpreting the past. This table also provided a relaxed setting for visitors and Hopi people to meet and talk to an "archaeologist" (Fig. 14.2). Many visitors to Homol'ovi have preconceived ideas about what archaeologists do. Interacting with the HUROP students helped challenge the stereotypes of archaeologists as adventurous treasure hunters and grave robbers.

By participating in the events at *Suvoyuki* Day, HUROP students also had the opportunity to interact with and learn from members of the Hopi Tribe. An important event at the beginning of Suvoyuki Day was a fun run. During the 3 years of the project, HUROP students helped with logistics (e.g., checking in runners, handing out water) or participated in the run. Before the start of the run, Hopi perspectives on running were shared. Hopi people traditionally run at sunrise to greet the day, pray for rain, and connect with the land and its resources. Running not only strengthens the individual, but benefits the community as a whole. For the *Suvoyuki* Day Run, official times were not recorded and no prizes were given to the fastest runners; instead, participants were greeted with shouts of thanks in Hopi as they finished. A HUROP student commented in his field journal that he was initially bothered by the fact that the run was not a competitive race. However, after participating, he viewed running in a different way, specifically that running did not always have to be about being faster or better than others.

Sharing information with the public about archaeology and Hopi traditions has always been a focus of the open house at the Homol'ovi State Park. Originally, archaeologists were the primary presenters of this information. Over the years, the event has become a joint effort with opportunities for visitors to interact with Hopi community members as well as archaeologists. This event is also important because it provides an opportunity for archaeologists, both students and professionals, to come together with Hopi people in a place where they have shared interests. Although Hopi people and archaeologists have a long history of interactions, the agreement between the Hopi Tribe and Arizona State Parks has created a new power dynamic and new opportunities for archaeologists and Hopi people to work together on interpretation and the events at the Park. In this time of state budget crises, these joint efforts have also become increasingly important for preserving the Park. When the Arizona state legislature discussed closing the Homol'ovi State Park in 2009, archaeologists, including HUROP students, representatives of the Hopi Tribe, and Arizona residents protested. Although these efforts did not keep the Park from closing, State Parks agreed to keep one ranger in residence at Homol'ovi to protect the sites, and in March 2011, it reopened after the Hopi Tribe offered to share the costs of operating the Park. The energy and relationships that were fostered during *Suvoyuki* Day have helped to sustain and focus the efforts to preserve the archaeological sites at Homol'ovi.

14.3.2 Building a Corn Roast

The corn roast on *Suvoyuki* Day provided other opportunities for both collaborative and experiential learning opportunities. Hopi people traditionally roast corn in a large bell-shaped pit which is lined with rocks and has a vent on the side to circulate air within the pit. The roasting begins by lighting a fire in the pit and adding fuel until the sides of the pit turn white in color from the heat. Once the flames die down, the corn is added, the pit and the vent are sealed, and the corn cooks in the pit overnight. In past years, a large hole was dug near the visitor center for the corn roast and then filled in after the roast to avoid erosion and injuries from falls. Although

Fig. 14.3 HUROP students and Hopi community members digging the corn roasting pit

this method worked, several Hopi people commented that the corn would taste better if roasted in a bell-shaped pit. I offered to have HUROP students help with the construction of the pit, if we could arrange to have a Hopi community member supervise the project.

In the summer of 2008, Susan Secakuku asked a Hopi elder to work with two HUROP students and the Homol'ovi intern from the Hopi Tribe to construct a corn roasting pit in an area near the visitor center that was used for experimental archaeology projects (Fig. 14.3). Funding for this project came from a grant received by the Hopi Tribe for interpretive activities at Homol'ovi. During the course of a hard day of digging, the pit was finished and the inside was left exposed to dry out. Every HUROP student assisted during the day of the roast. They gathered grease wood to use in the roasting, helped seal the pit, and removed the cooked corn from the pit the next morning. In the process of working together, Hopi community members shared aspects of their culture. For example, to celebrate the successful roast it is a Hopi tradition that all share the first ear of corn removed from the pit. As visitors gathered around to watch the corn come out of the pit, the HUROP students explained their role in the roast. This project, more than any other HUROP activity, captured the spirit of *Suvoyuki*. Although the creation of the pit and the satisfaction of producing a successful roast were wonderful, what was most important was the joint effort – the process of working together.

This pit was not only an important collaborative activity, but it also was an excellent opportunity to use a living tradition to gain insights into the archaeological record. One of the students who worked on the pit construction recorded the steps in the construction and the use of the pit for her research report that was part of the HUROP students' educational experience. Her paper described the excavation process, pit dimensions before and after the roasting due to portions of the sides that collapsed during the roast, changes in soil color after the roast, and the charred remnants of the roast. This information and subsequent excavation of the pit fill has provided insights into the formation of the layers within bell-shaped pits and the archaeological signatures of roasting activities.

14.3.3 Developing Exhibits Cases

Over the last two decades, collaboration with communities for exhibition development has been an increasingly common practice in museums (Phillips 2003). Using these collaborative models as an inspiration, I suggested that HUROP students help revise exhibits within the Homol'ovi State Park visitor center. This process began with students evaluating the existing exhibits and making suggestions about how to improve them, based on their experiences in exhibits at nearby archaeological parks such as the newly revised exhibits at Wupatki National Monument. Students noticed that the exhibits at Homol'ovi were uninteresting culture history presented in an abstract curatorial voice, or collections of objects with very little explanation. None of the signs included the voices of Hopi people or the archaeologists who worked at the Park. The only medium where a visitor could learn about the archaeological research at the Park was through an interactive computer display.

Developing new displays that informed the visitor about what archaeologists learn from objects was fairly easy. Students identified topics and archaeological objects for exhibits and then wrote text describing the archaeological interpretive process. The challenge was finding topics or objects where archaeological and Hopi perspectives could be discussed together.

After an evening lecture to HUROP students, a Hopi archaeologist mentioned that he would like to see some of the *piiki* stones found at ancestral Hopi villages at Homol'ovi put on display. *Piiki* is a paper-thin flat bread and an important traditional Hopi food (Kavena 1980). Archaeologists found the stones and uniquely shaped hearths used to make *piiki* in special structures near the plazas at the Homol'ovi pueblos (Adams 2002). Consequently, *piiki* made an ideal topic for an exhibit that included both archaeological and Hopi perspectives.

The process used to develop this exhibit was at times haphazard, but ended up being very important both as an educational experience for the students and for developing a conceptual framework for future exhibits. The process involved HUROP students developing the content for the exhibit using published materials, as well as discussions with archaeologists and Hopi women. Susan Secakuku and I edited the content that students wrote. For the 2007 *Suvoyuki* Day, a draft of the exhibit text was installed in an exhibit case with two archaeological *piiki* stones and a *piiki* bowl made by a contemporary potter. The student who wrote the text asked visitors for their impressions and revised the text based on these suggestions. During the next summer (2008), an intern from the University of Michigan Museum Studies Program helped to design the exhibit layout, using the content written by the students and a picture of the *piiki*-making demonstrations at *Suvoyuki* Day. After another round of comments from the archaeologists who had researched *piiki*-making at Homol'ovi and Susan Secakuku, we installed the completed text panel for *Suvoyuki* Day 2009.

Although it took a long time to complete a single exhibit panel, the process of developing this exhibit proved very important. The *piiki* exhibit helped focus subsequent discussions on the importance of presenting Hopi people's and archaeologists' perspectives. The student involvement also facilitated knowledge sharing between archaeologists and Hopi community members. For example, one of the Hopi interns at Homol'ovi told her mother that the HUROP students were working an exhibit that would include archaeological *piiki* stones. The mother was very excited about the exhibit but was puzzled by the fact that the *piiki* stones were broken and left behind at the site. Contemporary Hopi woman consider their *piiki* stone a family member and would never discarded it. This comment provided a very different perspective on these tools and raised a series of questions about the differences between the past and present that the HUROP students, the Hopi intern, and I discussed. I knew that the *piiki* was a Hopi food, but had not really understood its importance as a living tradition until this opportunity to learn about it with my students.

14.4   Reflections on Interpretive Collaboration in a Field School Setting

Forty-two students participated in HUROP between 2006 and 2008. One of the field school learning objectives was for students to gain a deeper understand of how archaeologists interpret the past by participating in fieldwork and then sharing what they learned about this process with the public. The student evaluations at the conclusion of each field season suggested that these educational goals had been met. Students were asked to assess the program and specifically to discuss whether they felt the integration of archaeological field work with public outreach was a valuable experience. All the student participants wrote highly favorable comments about this aspect of the project, and their subsequent career choices highlight this success. Over three-quarters of the students who participated in HUROP expanded on their interest in archaeology, museum studies, or heritage management through attending graduate school, finding a job, or pursing an internship in these fields. I have been able to keep in touch with many HUROP students, and several have commented that their experience on this field school prompted them to think about the importance of communicating research to the public and working with communities in new and different ways.

One disappointment of HUROP was that college students from the nearby reservations did not apply for the paid internships offered as part of HUROP, despite my contacts with individuals in the cultural resource management and heritage preservation programs of all of these tribes as part of the consultation process required for the archaeological excavations. Each year, recruiting materials were sent to these individuals and offices. Students with Native heritage did participate in the program, but they were not from the reservations around Homol'ovi. Difficulties attracting undergraduate students who grew up on reservations to apply to an archaeological field school are not unique to HUROP (Mills et al. 2008). In contrast, the Hopi Tribe was able to hire three summer interns to work at Homol'ovi. These internships, however, focused on opportunities to learn about the management of the Park and tourism. One of the reasons that Hopi college students were attracted to Hopi tribe's internship opportunity is the fact that it provided them with skills that were directly relevant to future jobs. Although they did not participate in the archaeological fieldwork, the Hopi interns were a vital part of the HUROP program. The interns lived and ate with us, making friends with many of the field school students; they also were interested in what we were finding and regularly came to visit our excavations.

In the United States, archaeologists have created scholarships for Native students, such as the Society for American Archaeology Native American Scholarship Fund, to participate in archaeological field and research opportunities. These resources are very important, but archaeologists need to also think about opportunities for Native college students to learn about archaeology that do not necessarily involve actually doing it. The aspects of HUROP in which the Hopi interns expressed the most interest were the ways in which their own culture was presented to the public and what archaeologists have learned about their history. They expressed little interest in collecting archaeological data.

The examples I discussed in this chapter also illustrate how a field school is an important setting for interpretive collaborations. Presenting both Native and archaeological perspectives is becoming increasingly common in the visitor centers at archaeological sites in the United States. However, the process of developing displays is often coordinated by the agency managing the site. At Homol'ovi, the student projects and educational goals of HUROP provided the venue for archaeologists and Hopi people to work directly with one another on interpretative projects.

Integrating public outreach and community collaborations into a field school curriculum was incredibly rewarding but also challenging. Coordinating the various aspects of the program took months of advanced planning. Moreover, HUROP would not have been feasible without the funding resources and a network of people who were willing to help. For example, each of the faculty mentors contributed to the educational component of the project and often helped with field logistics, student research, or the coordination of public outreach projects. My field crews also included three to five graduate students who were responsible for aspects of the field work, laboratory work, camp life, or interpretive projects. Many of the challenges of field work can be overcome with a good crew. Directing the Homol'ovi Undergraduate Research Opportunities Program also taught me how essential team work is for public outreach, community collaborations, and undergraduate education.

Acknowledgments Funding for the Homol'ovi Undergraduate Research Opportunities Program was provided by the National Science Foundation's Research Experiences for Undergraduates and the University of Michigan. HUROP would not have been possible without the support of Arizona State Parks. HUROP would not have been possible without the support of Arizona State Parks. Special thanks to all who participated in the program. You helped make Homol'ovi a more interesting place for visitors.

References

Adams, E. C. (2000). Homol'ovi: An ancestral Hopi place. *Archaeology Southwest, 14*(4), 1–4.

Adams, E. C. (2002). *Homol'ovi: An ancient Hopi settlement cluster.* Tucson: The University of Arizona Press.

Chilton, E. S. (2006). From the ground up: The effects of consultation on archaeological methods. In J. E. Kerber (Ed.), *Cross-cultural collaboration: Native peoples and archaeology in the Northeastern United States* (pp. 281–294). Lincoln: University of Nebraska.

Colwell-Chanthaphonh, C., & Ferguson, T. J. (2008). Introduction: The Collaborative continuum. In C. Colwell-Chanthaphonh & T. J. Ferguson (Eds.), *Collaboration in archaeological practice: Engaging descendent communities* (pp. 1–32). Lanham: AltaMira.

Ferguson, T. J., Dongoske, K. E., Yeatts, M., & Kuwanwisiwma, L. J. (2000). Hopi oral history and archaeology. In K. E. Dongoske, M. Aldenderfer, & K. Doehner (Eds.), *Working together: Native Americans and archaeologists* (pp. 45–60). Washington, DC: The Society for American Archaeology.

Fewkes, J. W. (1898). Preliminary account of an expedition to the Pueblo Ruins Near Winslow, Arizona in 1896. In *Bureau of American ethnology, Annual Report for 1896*. Government Printing Office, Washington, DC.

Howard, K. L., & Pardue, D. F. (1996). *Inventing the Southwest: The Fred Harvey Company and Native American Art*. Flagstaff: Northland.

Kavena, J. T. (1980). *Hopi cookery*. Tuscon: University of Arizona Press.

Labelle, J. M. (2003). Coffee cans and Folsom points: Why we cannot continue to ignore the artifact collectors. In L. J. Zimmerman, K. D. Vitelli, & J. Hollowell-Zimmer (Eds.), *Ethical Issues in archaeology* (pp. 15–127). Walnut Creek: AltaMira.

Mills, B. J., Altaha, M., Welch, J. R., & Ferguson, T. J. (2008). Field schools without trowels: Teaching archaeological ethics and heritage preservation in a collaborative context. In S. W. Silliman (Ed.), *Collaboration at the trowel's edge: Teaching and learning in indigenous archaeology* (Amerind Studies in Archaeology, pp. 25–49). Tucson: University of Arizona Press.

Phillips, R. B. (2003). Introduction. In L. Peers & A. Brown (Eds.), *Museums and source communities: A Routledge reader* (pp. 155–170). London: Routledge.

Rossen, J. (2008). Field school archaeology, activism, and politics in the Cayuga Homeland of Central New York. In S. W. Silliman (Ed.), *Collaboration at the trowel's edge: Teaching and learning in indigenous archaeology* (pp. 103–120). Tucson: University of Arizona Press.

Sekaquaptewa, E., Black, M. E., Malotki, E., & Hill, K. C. (1998). *Hopi Dictionary: Hopiikwa Lavaytutuveni: A Hopi-English Dictionary of the Third Mesa Dialect with an English-Hopi Finder List and a Sketch of Hopi Grammar*. Tucson: University of Arizona Press.

Silliman, S. W. (2008). Collaborative indigenous archaeology: Troweling at the edges, eyeing the center. In S. W. Silliman (Ed.), *Collaboration at the trowel's edge: Teaching and learning in indigenous archaeology* (pp. 1–24). Tucson: University of Arizona Press.

Silliman, S. W., & Sebastian Dring, K. H. (2008). Working on pasts for futures: Eastern Pequot Field School Archaeology in Connecticut. In S. W. Silliman (Ed.), *Collaboration at the trowel's edge: Teaching and learning in indigenous archaeology* (pp. 67–87). Tucson: University of Arizona Press.

Singleton, T. A., & Orser, C. E. (2003). Descendant communities: Linking people in the present to the past. In L. J. Zimmerman, K. D. Vitelli, & J. Hollowell-Zimmer (Eds.), *Ethical issues in archaeology* (pp. 143–152). Walnut Creek: AltaMira.

Walker, W. H. (2000). Ancient ritual. *Archaeology southwest, 14*(4), 8.

Chapter 15
Field Schools: People, Places, and Things in the Present

Harold Mytum

15.1 Introduction

Unlike much academic research archaeological fieldwork is normally a communal activity, and unlike science laboratory collaborative involvement, it can include people with little prior experience, who often live as well as work in close proximity. In addition, the physical settings in which fieldwork takes place also provide physical and emotional challenges. Most field schools take place away from the normal domestic arrangements for all participants, creating a particular social environment in which people discover about themselves and others. They also learn about archaeology, both through the deliberate training provided and through the incidental experiences gained by doing real-life archaeology in a research context.

The structure of field schools is now more coherent and planned than used to be the case, as the reflective studies in this book demonstrate, but it may be also instructive to see how the situation has changed in the last 40 years.

15.2 Field Training: An Autobiographical Interlude

In Britain during the 1970s, when I began my archaeological field experience, there were many opportunities for people like me who were still at high school to take part in research and contract archaeology projects where training was acquired in an *ad hoc* manner as different tasks were undertaken. An American graduate student and now a well-known historical archaeologist, Eric Klingelhofer, suffered my

H. Mytum (✉)
Centre for Manx Studies, School of Archaeology, Classics, and Egyptology,
University of Liverpool, Liverpool, UK
e-mail: h.mytum@liv.ac.uk

H. Mytum (ed.), *Global Perspectives on Archaeological Field Schools:*
Constructions of Knowledge and Experience, DOI 10.1007/978-1-4614-0433-0_15,
© Springer Science+Business Media, LLC 2012

assistance on one of his excavations in the historic town of Warwick, where I had already assisted on several other excavations directed by other archaeologists linked to Warwickshire County Museum. He kindly encouraged the Museum to sponsor me to attend a formal archaeological training excavation. The following year I attended the Wroxeter training excavation, staying at the slightly decaying splendor of Attingham Park, a stately home turned into an educational center. My place was sponsored by the museum, and the field school was directed by the late Graham Webster, doyenne of Roman military archaeology (1969a, 1980) and author of a standard work on archaeological excavation (Webster 1965). He already knew me, having taught number of evening classes on archaeology which I had attended every week over the previous couple of winters. The experience was not perhaps what Eric Klingelhofer has imagined. While Graham Webster taught me a great deal about Roman pottery (Webster 1969b), and about the complexities of stratigraphy and how to "read" excavation reports, all of which was of great value in my subsequent career, my field training was less structured than might be expected in such a context.

The field school was taking place in the center of the Roman town of Wroxeter, now a green-field site not dissimilar in character and survival to that of Silchester (Chap. 4). Lying adjacent to the already-excavated bath buildings which were on display to the public, the particular site being investigated had already been investigated through trenching by a previous generation of archaeologists led by Kathleen Kenyon (1937). Graham Webster gave regular morning introductory lectures, but he generously instructed me to avoid these, as they would only cover what he had already taught me in evening classes, so I was left to excavate alone during these times.

My main task was to empty, as rapidly as possible, the backfilled trenches excavated previously, so that the sequence of stratigraphy in the trench sides could be examined, and the lowest deposits that may not have been fully investigated could be exposed. Graham Webster's main research interest was in these earliest layers as he was concerned with rewriting the story of the origins of the town and the possible military phase at its inception. To the background of Webster's distinctive droning voice emanating from the wooden shed in which the rest of the class was closeted, I worked alone using pick and shovel to loosen the spoil within the trench, and then shovel it over my shoulder into the wheelbarrow which I periodically emptied by pushing it up a ramp past the shed onto the large mound of soil accumulated over several seasons. Watched with amusement by prisoners and guards on limited day release from an open prison some miles away, and who were meant to be helping with the spoil removal but never did any work at all, I did wonder what I would gain from this supposed training excavation. In fact I learnt a great deal.

Instead of slowly and carefully excavating deposits within Roman buildings, my exposure of the trench sides revealed complex stratigraphy revealing hundreds of years of development on the site, and this encouraged full appreciation of sequence which my previous experience had not developed to such a degree. Moreover, opening these old trenches physically recreated the style of the 1930s excavation, to contrast with the Wheelerian box-style excavation of Graham Webster, already old fashioned by the 1970s. Moreover, beyond the fence surrounding our excavations, in the distance and just within earshot was the methodologically cutting-edge open-area excavations of Philip Barker (1977), dealing in quite a different way with an area of the town also dissected by Kenyon's trenches and revealing phases of the later settlement (Barker et al. 1977) that Webster's methods failed to identify, as had been the case with Kenyon's. Being able to compare and contrast such a range of methodologies, and to discuss them with those still holding firm to the traditional Wheelerian methods, was instructive, even though I was personally excited by the open-area approach which I implemented when appropriate later in my career.

My efforts contributed to Webster's understanding of the early military history, though the final report only appeared after his death in 2001 (Webster and Chadderton 2002). But I also gained a great deal by using my existing knowledge, drawing upon the expertise of others, and comparing Kenyon's, Webster's, and Barker's excavation methods all on the same site. Thus, some of what I learnt was as planned within the training program, but much came indirectly, and this is often the case. While contemporary pedagogic practice is much more robust, and all students will receive more coherent training than I received, what is important is the attitude towards teaching and learning of both students and staff. Graham Webster wanted me to learn and provided me with solid digging experience rather than a patchy experience interspersed by lectures that would have only reprised his earlier lecturing. His staff, including Don Macreth, also gave insights on excavation technique and interpretation that was appropriate to my level of experience, asking me questions that they would not have asked a real novice. Though not phrased in the newly coined pedagogic rhetoric of the twenty-first century, they knew how to teach, inspire, and encourage. While the content and style of the training excavation was undoubtedly not that imagined by Eric Klingelhofer, I am indebted to his initiative, and Warwick Museum's generosity, as that helped direct my own archaeological practice in new ways, building on the strengths of the older traditions.

The field school, because of its earthy reality, can escape from educationalists' jargon to be a physical and intellectual experience but which can be one informed by that background. I learnt more on the Wroxeter training excavation than I had expected after my first day, and a wider range of insights into archaeological method than my teachers had anticipated, because I engaged with the experience. I have continued to do this every season, with whatever responsibility I have held, and the same should be the case with all who set out for the field today. A passive attitude by students, or a formulaic approach by staff, will limit the potential of the experience and, although the technical learning outcomes may be achieved, a chance for personal and professional growth can be missed. An overemphasis by students on achieving high grades, on competition, and on ingratiation with staff will lead to alienation and frustration, and the missing of many opportunities that come by chance, that are part of the rich complexity of field school life. Likewise, staff who do not balance the ethical and research requirements with those of teaching in an effort to satisfy the potentially infinite demands of students will fail to reveal the complexity of field projects to the students and limit their learning of self-reliance and independent thought. This domination by overindulgence might reduce the chance of critical comment by students, but field school directors and their staff should be willing to engage in critical discussions and accept that differences of opinion are acceptable. Given that the director has the wider responsibility for the project and its outcomes of all kinds, that view should prevail, but not without siren voices being recognized, as long as these are not disruptive. At all times, differences in interpretation or theoretical position should be differentiated from personal feelings or challenges to the established structures of power and responsibility.

15.3   Field Schools Today and Tomorrow

Field schools vary in the levels of technology applied, the complexity of research questions asked, and the size of work force and number of collaborative partners. As archaeology has become more scientific in some regards, but also more aware of its place as action within a cultural context of the present, so the range of theoretical and methodological approaches that might be applied has multiplied. The result is that field school subcultures vary even more greatly than ever before, but the explicit reflexivity of the experience is now more easily transferred to new situations than was previously the case.

Potential students and staff should remember that they are to enter into a partially closed world where the physical, intellectual, and social concerns of the field school become engrossing. Even those field schools that consist completely or largely of students from one university, being led by a teacher by whom they have already been taught, will create dynamics quite different from those experienced in class. Particularly in those field schools that are physically isolated, and where food and transport are all provided, a form of enclosed lifestyle develops which is largely comforting and supporting, though for a small minority may be claustrophobic. While staff and students may be able to create and maintain a certain image during weekly classes, faced with long periods of close contact, periods of fatigue, and moments of frustration, such masks will inevitably drop. Those who enter into such endeavors openly have a much less stressful initiation and still learn much about themselves; those with a self-image that is challenged find this aspect of the field school far more unsettling than the new experiences of learning archaeological field methods. People's backgrounds can be at least in part left behind during the field school, successes and failures unknown and irrelevant, but may come back to the fore once the experience is over.

Staff and students need to be aware of the complex personal dynamics that occur on field schools and should be ready to act to prevent any development of ill feeling between members. Staff should be sensitive to interpersonal tensions and diffuse these by moving personnel about the site, and if necessary, rearrange domestic arrangements. Sometimes a disaffected student, usually discovering the fieldwork or the living conditions are not to their liking, can create an atmosphere of dissatisfaction. It is essential that any such opinion should be evaluated, and if there are justifiable causes, they should be resolved. This is rarely the case, however, and such students need to be managed to contain their disillusion and, with time and

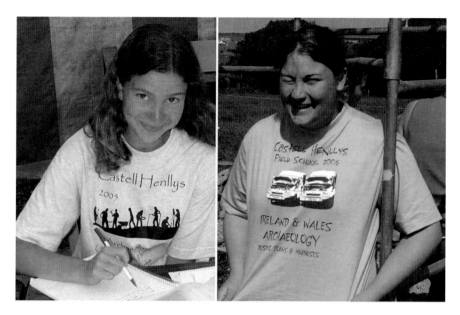

Fig. 15.1 Field school T-shirts both reflect and create group identity during the project and create memories and show participation in a rite of passage subsequently

effort, they may be brought round to accepting and even enjoying the experience, even if it would never be repeated. Peer pressure can work for good or ill on a field school, and once momentum shifts in a certain direction, it can be difficult to redirect it. Staff should be sensitive throughout to such mood swings, as it may occur at the beginning, when many may feel isolated and vulnerable, later in the field school as fatigue and minor irritations with the schedule, conditions, or others' habits can build up, or towards the end when completion of project and assignment tasks conflict with the preparations for entering back into the wider world beyond the field school. The careful, slight adjustment of arrangements, a quiet encouragement or even warning, and morale-boosting surprise activity or enthusiastic summary of recent achievements can almost always have a dramatic effect. Making very slight adjustments can mean that it appears to many of the participants that there is no person management and that everything runs smoothly by chance – or is inevitable. This is far from the case; field school directors are above all facilitators of people (both students and staff); it is through them that the teaching, learning, and research results all flow.

Friendships forged during the field school may last beyond its remit, though sometimes they are never quite the same in different situations. Students may scatter across large areas back to their homes and universities, though now many maintain contact through social network sites and email. The popularity of field school T-shirts reveals the sense of belonging and power of the experience enjoyed by students (Fig. 15.1). Long-running field schools provide a shared experience across

Fig. 15.2 This excavation crew photograph reflects the field school ethos of people and place bonded through shared experiences

cohorts of students, a bond that can provide an opportunity to make contact and to enhance career prospects; networking within archaeology can begin on the field school. Field school directors are often requested to provide letters of support and references for students applying for graduate school or CRM posts, and assistant field school staff also gain valuable experience that can be confirmed by an experienced leader. Field school directors often know more about the strengths and weaknesses in character, applied intellect, and employment skills of a student than any other member of university staff.

The field school forms a web of interaction between theory and practice, past and present, teacher and taught, archaeology and community, people and things, and people and places. For different members of the team, the importance of these complex interactions will vary, and the director's perceptions and experiences will be very different to those of the novice fieldworker. All, however, are valid, and all are necessary to create an effective team which achieves the pedagogic and research goals. The standard, familiar, field school group photograph may reveal to the insider friendships and alliances, hierarchies, and tensions, but behind the smiles lies a shared endeavor which should not be underestimated (Fig. 15.2). Often with a clear reference to place, and through their hair and clothes fashions situated in time, these images create, recall, and recreate memories of fieldwork times past and, like anonymous family photographs, may hold little significance to the casual observer.

However, to those for whom these images are made, they encapsulate a forever-memorable time in their lives, a unique set of experiences which will not be repeated even if further field schools are undertaken.

While the internal dynamic of the field school is almost always strong, some field schools are very much part of a wider world, through their location or the ways in which they are designed to link with local communities, other heritage professionals, or other projects. Unlike some outdoor experiences, that are largely of and for themselves as experiences, field schools have additional obligations to others. Whatever the public outreach mission, they will create data that will be of interest to others and will require ordering, interpretation, and curation. Field school staff have obligations to publish or otherwise disseminate findings, not only to past field school participants, but also to the wider archaeological community. Contract archaeology now often carries out the largest-scale field projects in archaeology today, but the long-term research excavation, often supported at least in part by a field school, still has a special place in archaeology. These projects allow a more measured, reflective, and potentially experimental research design that can make a distinctive contribution to the development of the discipline, as well as train the next generation of practitioners. Despite financial pressures, field schools still flourish. Those who have written for this book trust that their accumulated experiences and insights will help those embarking on this special part of the archaeologist's life gain the most from their experience. For those already initiated, we wish that these comparative commentaries can lead to new ideas, strategies, and ambitions, so that field schools become more robust and effective, and can continue to inspire future generations.

References

Barker, P. A. (1977). *Techniques of archaeological excavation.* New York: Universe Books.

Barker, P. A., White, R., Pretty, K., Bird, H., & Corbishley, M. (1977). *The Baths Basilica Wroxeter. Excavations 1966-90.* English Heritage Archaeological Report 8. London: English Heritage.

Kenyon, K. M. (1937). *Excavations at Viroconium, 1936-1937.* Shrewsbury: Shropshire Archaeological Society.

Webster, G. (1965). *Practical archaeology: An introduction to archaeological field-work and excavation.* London: Batsford.

Webster, G. (1969a). *The Roman Imperial Army of the first and second centuries A.D..* London: Black.

Webster, G. (Ed.). (1969b). *Romano-British coarse pottery: A students' guide.* London: Council for British Archaeology.

Webster, G. (1980). *The Roman Invasion of Britain.* London: Batsford.

Webster, G., & Chadderton, J. (Eds.) (2002). *The Legionary Fortress at Wroxeter: Excavations by Graham Webster, 1955–85.* English Heritage Archaeological Report 19. London: English Heritage.

Index

H. Mytum (ed.), *Global Perspectives on Archaeological Field Schools:*
Constructions of Knowledge and Experience, DOI 10.1007/978-1-4614-0433-0,
© Springer Science+Business Media, LLC 2012

LIBRARY, UNIVERSITY OF CHESTER

MIX
Papier aus verantwortungsvollen Quellen
Paper from responsible sources
FSC® C105338

Printed by Books on Demand, Germany